CAMBRIDGE LIBRARY COLLECTION

Books of enduring scholarly value

Life Sciences

Until the nineteenth century, the various subjects now known as the life sciences were regarded either as arcane studies which had little impact on ordinary daily life, or as a genteel hobby for the leisured classes. The increasing academic rigour and systematisation brought to the study of botany, zoology and other disciplines, and their adoption in university curricula, are reflected in the books reissued in this series.

Life and Letters of Thomas Henry Huxley

Thomas Henry Huxley (1825–95), the English biologist and naturalist, was known as 'Darwin's Bulldog', and is best remembered today for his vociferous support for Darwin's theory of evolution. He was, however, an influential naturalist, anatomist and religious thinker, who coined the term 'agnostic' to describe his own beliefs. Almost entirely self-educated, he became an authority in anatomy and palaeontology, and after the discovery of the archaeopteryx, he was the first to suggest that birds had evolved from dinosaurs. He was also a keen promoter of scientific education who strove to make science a paid profession, not dependent on patronage or wealth. Published in 1903, this three-volume work, compiled by his son Leonard Huxley, is the second and most complete edition of Huxley's biography and selected letters. Volume 2 covers the period 1870–86, including Huxley's American lecture tour, and the death of his friend Charles Darwin in 1882.

T0188080

Cambridge University Press has long been a pioneer in the reissuing of out-of-print titles from its own backlist, producing digital reprints of books that are still sought after by scholars and students but could not be reprinted economically using traditional technology. The Cambridge Library Collection extends this activity to a wider range of books which are still of importance to researchers and professionals, either for the source material they contain, or as landmarks in the history of their academic discipline.

Drawing from the world-renowned collections in the Cambridge University Library, and guided by the advice of experts in each subject area, Cambridge University Press is using state-of-the-art scanning machines in its own Printing House to capture the content of each book selected for inclusion. The files are processed to give a consistently clear, crisp image, and the books finished to the high quality standard for which the Press is recognised around the world. The latest print-on-demand technology ensures that the books will remain available indefinitely, and that orders for single or multiple copies can quickly be supplied.

The Cambridge Library Collection brings back to life books of enduring scholarly value (including out-of-copyright works originally issued by other publishers) across a wide range of disciplines in the humanities and social sciences and in science and technology.

Life and Letters of Thomas Henry Huxley

VOLUME 2

LEONARD HUXLEY

CAMBRIDGE
UNIVERSITY PRESS

CAMBRIDGE UNIVERSITY PRESS

Cambridge, New York, Melbourne, Madrid, Cape Town,
Singapore, São Paolo, Delhi, Tokyo, Mexico City

Published in the United States of America by Cambridge University Press, New York

www.cambridge.org
Information on this title: www.cambridge.org/9781108040464

© in this compilation Cambridge University Press 2012

This edition first published 1903
This digitally printed version 2012

ISBN 978-1-108-04046-4 Paperback

LIFE AND LETTERS

OF

THOMAS HENRY HUXLEY

Maull & Fox photo.   Walker & Cockerell ph. sc

1857.

Life and Letters

OF

Thomas Henry Huxley

BY HIS SON

LEONARD HUXLEY

IN THREE VOLUMES

VOL. II

London

MACMILLAN AND CO., Limited

NEW YORK : THE MACMILLAN COMPANY

1903

First Edition, 2 vols. 8vo, October 1900.
Reprinted November and December 1900.
Second Edition, Globe 8vo (Eversley Series, 3 vols.), 1903.

CONTENTS

v

PORTRAIT

T. H. HUXLEY

From a Photograph by Maull and Polyblank, 1857.

Frontispiece.

CHAPTER I

1870

WITH the year 1870 comes another turning-point in Huxley's career. From his return to England in 1850 till 1854 he had endured four years of hard struggle, of hope deferred ; his reputation as a zoologist had been established before his arrival, and was more than confirmed by his personal energy and power. When at length settled in the professorship at Jermyn Street, he was so far from thinking himself more than a beginner who had learned to work in one corner of the field of knowledge, still needing deep research into all kindred subjects in order to know the true bearings of his own little portion, that he treated the next six years simply as years of further apprenticeship. Under the suggestive power of the *Origin of Species* all these scattered studies fell suddenly into due rank and order; the philosophic unity he had so long been seeking inspired his thought with tenfold vigour, and the battle at

Oxford in defence of the new hypothesis first brought him before the public eye as one who not only had the courage of his convictions when attacked, but could, and more, would, carry the war effectively into the enemy's country. And for the next ten years he was commonly identified with the championship of the most unpopular view of the time; a fighter, an assailant of long-established fallacies, he was too often considered a mere iconoclast, a subverter of every other well-rooted institution, theological, educational, or moral.

It is difficult now to realise with what feelings he was regarded in the average respectable household in the sixties and early seventies. His name was anathema; he was a terrible example of intellectual pravity beyond redemption, a man with opinions such as cannot be held "without grave personal sin on his part" (as was once said of Mill by W. G. Ward, see p. 142), the representative in his single person of rationalism, materialism, atheism, or if there be any more abhorrent "ism"—in token of which as late as 1892 an absurd zealot at the headquarters of the Salvation Army crowned an abusive letter to him at Eastbourne by the statement, "I hear you have a local reputation as a Bradlaughite."

But now official life began to lay closer hold upon him. He came forward also as a leader in the struggle for educational reform, seeking not only to perfect his own biological teaching, but to show, in theory and practice, how scientific training might be introduced into the general system of education. He was more than once asked to stand for Parliament, but refused, thinking he could do more useful work for his country outside.

The publication in 1870 of *Lay Sermons*, the first of a series of similar volumes, served, by concentrating his moral and intellectual philosophy, to make his influence as a teacher of men more widely felt. The "active scepticism," whose conclusions many feared, was yet acknowledged as the quality of mind which had made him one of the clearest thinkers and safest scientific guides of his time, while his keen sense of right and wrong made the more reflective of those who opposed his conclusions hesitate long before expressing a doubt as to the good influence of his writings. This view is very clearly expressed in a review of the book in the *Nation* (New York, 1870, xi. 407).

And as another review of the *Lay Sermons* puts it (*Nature*, iii. 22), he began to be made a kind of popular oracle, yet refused to prophesy smooth things.

During the earlier period, with more public demands made upon him than upon most men of science of his age and standing, with the burden of four Royal Commissions and increasing work in learned societies in addition to his regular lecturing and official paleontological work, and the many addresses and discourses in which he spread abroad in the popular mind the leaven of new ideas upon nature and education and the progress of thought, he was still constantly at work on biological researches of his own, many of which took shape in the Hunterian lectures at the College of Surgeons from 1863-1870. But from 1870 onward, the time he could spare to such research grew less and less. For eight years he was continuously on one Royal Commission after another. His administrative work on learned societies continued to increase; in 1869-70 he held the presidency of the Ethnological Society, with a view to effecting the amalgamation with the Anthropological, "the plan," as he calls it, "for uniting the Societies which occupy themselves with man (that excludes 'Society' which occupies itself chiefly with woman)." He became President of the Geological Society in 1872, and for nearly ten years, from 1871 to 1880, he was secretary of the Royal Society, an office which occupied no small portion of his time and thought, "for he had formed a very high ideal of the duties of the Society as the head of science in this country, and was determined that it should not at least fall short through any lack of exertion on his part" (Sir M. Foster, R. S. Obit. Not.).[1]

The year 1870 itself was one of the busiest he had ever known. He published one biological and four paleontological memoirs, and sat on two Royal Commissions, one on the Contagious Diseases Acts, the other on Scientific Instruction, which continued until 1875.

[1] See Appendix II.

The three addresses which he gave in the autumn, and his election to the School Board will be spoken of later; in the first part of the year he read two papers at the Ethnological Society, of which he was President, on "The Geographical Distribution of the Chief Modifications of Mankind," March 9—and on "The Ethnology of Britain," May 10—the substance of which appeared in the *Contemporary Review* for July under the title of "Some Fixed Points in British Ethnology" (*Coll. Ess.* vii. 253). As President also of the Geological Society and of the British Association, he had two important addresses to deliver. In addition to this, he delivered an address before the Y.M.C.A at Cambridge on "Descartes' Discourse."

How busy he was may be gathered from his refusal of an invitation to Down :—

26 ABBEY PLACE, *Jan.* 21, 1870.

MY DEAR DARWIN—It is hard to resist an invitation of yours—but I dine out on Saturday; and next week three evenings are abolished by Societies of one kind or another. And there is that horrid Geological address looming in the future !

I am afraid I must deny myself at present.

I am glad you liked the sermon. Did you see the "Devonshire man's" attack in the *Pall Mall ?*

I have been wasting my time in polishing that worthy off. I would not have troubled myself about him, if it were not for the political bearing of the Celt question just now.

My wife sends her love to all you.—Ever yours,

T. H. HUXLEY.

The reference to the "Devonshire Man" is as follows:—Huxley had been speaking of the strong similarity between Gaul and German, Celt and Teuton, before the change of character brought about by the Latin conquest; and of the similar commixture, a dash of Anglo-Saxon in the mass of Celtic, which prevailed in our western borders and many parts of Ireland, *e.g.* Tipperary.

The "Devonshire Man" wrote on Jan. 18 to the *Pall Mall Gazette*, objecting to the statement that "Devonshire men are as little Anglo-Saxons as Northumbrians are Welsh." Huxley replied on the 21st, meeting his historical arguments with citations from Freeman, and especially by completing his opponent's quotation from Cæsar, to show that under certain conditions, the Gaul was indistinguishable from the German. The assertion that the Anglo-Saxon character is midway between the pure French or Irish and the Teutonic, he met with the previous question, Who is the pure Frenchman? Picard, Provençal, or Breton? or the pure Irish? Milesian, Firbolg, or Cruithneach?

But the "Devonshire Man" did not confine himself to science. He indulged in various personalities, to the smartest of which, a parody of Sydney Smith's dictum on Dr. Whewell, Huxley replied :—

" A Devonshire Man" is good enough to say of me that "cutting up monkeys is his forte, and cutting up men is his foible." With your permission, I propose to cut up " A Devonshire Man"; but I leave it to the public to judge whether, when so employed, my occupation is to be referred to the former or to the latter category.

For this he was roundly lectured by the *Spectator* on January 29, in an article under the heading "Pope Huxley." Regardless of the rights or wrongs of the controversy, he was chidden for the abusive language of the above paragraph, and told that he was a very good anatomist, but had better not enter into discussions on other subjects.

The same question is developed in the address to the Ethnological Society later in the year and in "Some Fixed Points in British Ethnology" (see above, p. 5), and reiterated in an address from the chair in Section D at the British Association in 1878 at Dublin, and in a letter to the *Times* for October 12, 1887, apropos of a leading article upon " British Race-types of To-day."

Letter-writing was difficult under such pressure of work, but the claims of absent friends were not wholly forgotten, though left on one side for a time, and the warm-hearted Dohrn, who could not bear to think himself forgotten, managed to get a letter out of him—not on scientific business.

26 ABBEY PLACE, Jun. 30, 1870.

MY DEAR DOHRN—In one sense I deserve all the hard things you may have said and thought about me, for it is really scandalous and indefensible that I have not written to you. But in another sense, I do not, for I have very often thought about you and your doings, and as I have told you once before, your memory always remains green in the " happy family."

But what between the incessant pressure of work and an inborn aversion to letter-writing, I become a worse and worse correspondent the longer I live, and unless I can find one or two friends who will [be] content to bear with my infirmities and believe that however long before we meet, I shall be ready to take them up again exactly where I left off, I shall be a friendless old man.

As for your old Goethe, you are mistaken. The Scripture says that "a living dog is better than a dead lion," and I am a living dog. By the way, I bought Cotta's edition of him the other day, and there he stands on my bookcase in all the glory of gilt, black, and marble edges. Do you know I did a version of his *Aphorisms on Nature* into English the other day.[1] It astonishes the British Philistines not a little. When they began to read it they thought it was mine, and that I had suddenly gone mad !

But to return to your affairs instead of my own. I received your volume on the *Arthropods* the other day, but I shall not be able to look at it for the next three weeks, as I am in the midst of my lectures, and have an annual address to deliver to the Geological Society on the 18th February, when, I am happy to say, my tenure of office as President expires.

After that I shall be only too glad to plunge into your doings and, as always, I shall follow your work with the heartiest interest. But I wish you would not take it into your head that Darwin or I, or any one else thinks otherwise than highly of you, or that you need "re-establishing" in any one's eyes. But I hope you will not have finished your work before the autumn, as they have made me President of the British Association this year, and I shall be very busy with my address in the summer. The meeting is to take place in Liverpool on the 14th September, and I live in hope that you will be

---

[1] For the first number of *Nature*, November 1869.

able to come over.   Let me know if you can, that I may
secure you good quarters.

I shall ask the wife to fill up the next half-sheet.
But for Heaven's sake don't be angry with me in English
again.   It's far worse than a scolding in Deutsch, and I
have as little forgotten my German as I have my German
friends.

On February 18 he delivered his farewell address [1]
to the Geological Society, on laying down the office
of President.   He took the opportunity to revise his
address to the Society in 1862, and pointed out the
growth of evidence in favour of the evolution theory,
and in particular traced the paleontological history of
the horse, through a series of fossil types approaching
more and more to a generalised ungulate type and
reaching back to a three-toed ancestor, or collateral
of such an ancestor, itself possessing rudiments of
the two other toes which appertain to the average
quadruped.

If (he said) the expectation raised by the splints of
the horses that, in some ancestor of the horses, these
splints would be found to be complete digits, has been
verified, we are furnished with very strong reasons for
looking for a no less complete verification of the expectation that the three-toed *Plagiolophus*-like "avus" of
the horse must have been a five-toed "atavus" at some
early period.

Six years afterwards, this forecast of paleon
tological research was to be fulfilled, but at the
expense of the European ancestry of the horse.   A

---

[1] "Paleontology and the Doctrine of Evolution," *Coll. Ess.* viii.

series of ancestors, similar to these European fossils, but still more equine, and extending in unbroken order much farther back in geological time, was discovered in America. His use of this in his New York lectures as demonstrative evidence of evolution, and the immediate fulfilment of a further prophecy of his will be told in due course.

His address to the Cambridge Y.M.C.A., "A Commentary on Descartes' 'Discourse touching the method of using reason rightly, and of seeking scientific truth,'" was delivered on March 24. This was an attempt to give this distinctively Christian audience some vision of the world of science and philosophy, which is neither Christian nor Un-christian, but Extra-christian, and to show "by what methods the dwellers therein try to distinguish truth from falsehood, in regard to some of the deepest and most difficult problems that beset humanity, "in order to be clear about their actions, and to walk sure-footedly in this life," as Descartes says. For Descartes had laid the foundation of his own guiding principle of "active scepticism, which strives to conquer itself."

Here again, as in the *Physical Basis of Life*, but with more detail, he explains how far materialism is legitimate, is, in fact, a sort of shorthand idealism. This essay, too, contains the often-quoted passage, apropos of the "introduction of Calvinism into science."

I protest that if some great Power would agree to make me always think what is true and do what is right, on condition of being turned into a sort of clock and wound up every morning before I got out of bed, I should instantly close with the offer.   The only freedom I care about is the freedom to do right; the freedom to do wrong I am ready to part with on the cheapest terms to any one who will take it of me.

This was the latest of the essays included in *Lay Sermons, Addresses and Reviews*, which came out, with a dedicatory letter to Tyndall, in the summer of 1870, and, whether on account of its subject matter or its title, always remained his most popular volume of essays.

To the same period belongs a letter to Matthew Arnold about his book *St. Paul and Protestantism.*

My dear Arnold—Many thanks for your book which I have been diving into at odd times as leisure served, and picking up many good things.

One of the best is what you say near the end about science gradually conquering the materialism of popular religion.

It will startle the Puritans who always coolly put the matter the other way ; but it is profoundly true.

These people are for the most part mere idolaters with a Bible-fetish, who urgently stand in need of conversion by Extra-christian Missionaries.

It takes all one's practical experience of the importance of Puritan ways of thinking to overcome one's feeling of the unreality of their beliefs.   I had pretty well forgotten how real to them " the man in the next street " is, till your citation of their horribly absurd dogmas reminded me of it.   If you can persuade them that Paul is fairly interpretable in your sense, it may be the beginning of better things, but I have my doubts if Paul would own you, if he could return to expound his own epistles.

I am glad you like my Descartes article. My business with my scientific friends is something like yours with the Puritans, nature being *our* Paul.—Ever yours very faithfully,   T. H. HUXLEY.

26 ABBEY PLAOE, *May* 10, 1870.

From the 14th to the 24th of April Huxley, accompanied by his friend Hooker, made a trip to the Eifel country. His sketch-book is full of rapid sketches of the country, many of them geological; one day indeed there are eight, another nine such.

Tyndall was invited to join the party, and at first accepted, but then recollected the preliminaries which had to be carried out before his lectures on electricity at the end of the month. So he writes on April 6 :—

ROYAL INSTITUTION, 6 *April.*

MY DEAR HUXLEY—I was rendered drunk by the excess of prospective pleasure when you mentioned the Eifel yesterday, and took no account of my lectures. They begin on the 28th, and I have studiously to this hour excluded them from my thought. I have made arrangements to see various experiments involving the practical application of electricity before the lectures begin ; I find myself, in short, cut off from the expedition. My regret on this score is commensurable with the pleasures I promised myself. Confound the lectures !

And yours [1] on Friday is creating a pretty hubbub already. I am torn to pieces by women in search of

---

[1] *On the Pedigree of the Horse*, April 8, 1870, which was never brought out in book form.

tickets. Anything that touches progenitorship interests them. You will have a crammed house, I doubt not.— Yours ever,   JOHN TYNDALL.

Huxley replied :—

GEOLOGICAL SURVEY OF ENGLAND AND WALES,
April 6, 1870.

MY DEAR TYNDALL—

DAMN

the

L

e

c

t

u

r

e

s.   T. H. H.

That's a practical application of electricity for you.

In June he writes to his wife, who has taken a sick child to the seaside :—

I hear a curious rumour (which is not for circulation), that Froude and I have been proposed for D.C.L.'s at Commemoration, and that the proposition has been bitterly and strongly opposed by Pusey.[1] They say there has been a regular row in Oxford about it. I suppose this is at the bottom of Jowett's not writing to me. But I hope that he won't fancy that I should be disgusted at the opposition and object to come [*i.e.* to pay his regular visit to Balliol]. On the contrary, the more complete Pusey's success, the more desirable it is that I should show my face there. Altogether it is an awkward position, as I am supposed to know nothing of what is going on.

---

[1] Huxley ultimately received his D.C.L. in 1885.

The situation is further developed in a letter to Darwin :—

JERMYN STREET, *June* 22, 1870.

MY DEAR DARWIN—I sent the books to Queen Anne St. this morning. Pray keep them as long as you like, as I am not using them.

I am greatly disgusted that you are coming up to London this week, as we shall be out of town next Sunday. It is the rarest thing in the world for us to be away, and you have pitched upon the one day. Cannot we arrange some other day ?

I wish you could have gone to Oxford, not for your sake, but for theirs. There seems to have been a tremendous shindy in the Hebdomadal board about certain persons who were proposed ; and I am told that Pusey came to London to ascertain from a trustworthy friend who were the blackest heretics out of the list proposed, and that he was glad to assent to your being doctored, when he got back, in order to keep out seven devils worse than that first !

Ever, oh Coryphaeus diabolicus, your faithful follower,

T. H. HUXLEY.

The choice of a subject for his Presidential Address at the British Association for 1870, a subject which, as he put it, "has lain chiefly in a land flowing with the abominable, and peopled with mere grubs and mouldiness," was suggested by a recent controversy upon the origin of life, in which the experiments of Dr. Bastian, then Professor of Pathological Anatomy at University College, London, which seemed to prove spontaneous generation, were shown by Professor Tyndall to contain a flaw. Huxley had naturally been deeply interested from the first; he had been consulted by Dr. Bastian, and, I believe, had advised him not to publish until he had made quite sure of his ground. This question and the preparation of the course of Elementary Biology [1] led him to carry on a series of investigations lasting over two years, which took shape in a paper upon "Penicillium, Torula, and Bacterium," [2] first read in Section D at the British Association, 1870 ; and in his article on "Yeast" in the *Contemporary Review* for December 1871. He laboriously repeated Pasteur's experiments, and for years a quantity of flasks and cultures used in this work remained at South Kensington, until they were destroyed in the eighties. Of this work Sir J. Hooker writes to him :—

> You have made an immense leap in the association of forms, and I cannot but suppose you approach the final solution. . . .
> I have always fancied that it was rather brains and boldness, than eyes or microscopes that the mycologists wanted, and that there was more brains in Berkeley's [3] crude discoveries than in the very best of the French and German microscopic verifications of them, who filch away the credit of them from under Berkeley's nose, and pooh-pooh his reasoning, but for which we should be, as we were.

In his Presidential Address, "Biogenesis and Abiogenesis" (*Coll. Ess.* viii. p. 229), he discussed the rival theories of spontaneous generation and the

---

[1] See p. 81, *sqq.*
[2] *Quart. Journ. Micr. Sci.*, 1870, x. pp. 355-362.
[3] Rev. M. J. Berkeley.

universal derivation of life from precedent life, and professed his belief, as an act of philosophic faith, that at some remote period, life had arisen out of inanimate matter, though there was no evidence that anything of the sort had occurred recently, the germ theory explaining many supposed cases of spontaneous generation.   The history of the subject, indeed, showed "the great tragedy of Science—the slaying of a beautiful hypothesis by an ugly fact—which is so constantly being enacted under the eyes of philosophers," and recalled the warning "that it is one thing to refute a proposition, and another to prove the truth of a doctrine which, implicitly or explicitly, contradicts that proposition."

Two letters to Dr. Dohrn refer to this address and to the meeting of the Association.

JERMYN STREET, *April* 30, 1870.

MY DEAR WHIRLWIND—I have received your two letters; and I was just revolving in my mind how best to meet your wishes in regard to the very important project mentioned in the first, when the second arrived and put me at rest.

I hope I need not say how heartily I enter into all your views, and how glad I shall be to see your plan for "Stations"[1] carried into effect.   Nothing could have a greater influence upon the progress of zoology.

A plan was set afoot here some time ago to establish a great marine Aquarium at Brighton by means of a company.   They asked me to be their President, but I declined, on the ground that I did not desire to become

---

[1] Dr. Dohrn succeeded in establishing such a zoological "station" at Naples.

connected with any commercial undertaking. What has become of the scheme I do not know, but I doubt whether it would be of any use to you, even if any connection could be established.

As soon as you have any statement of your project ready, send it to me and I will take care that it is brought prominently before the British public so as to stir up their minds. And then we will have a regular field-day about it in Section D at Liverpool.

Let me know your new ideas about insects and vertebrata as soon as possible, and I promise to do my best to pull them to pieces. What between Kowalewsky and his Ascidians, Miklucho-Maclay and his Fish-brains, and you and your Arthropods, I am becoming schwindel-suchtig, and spend my time mainly in that pious ejacula-tion "Donner und Blitz," in which, as you know, I seek relief. Then there is our Bastian who is making living things by the following combination :—

> R   Ammoniae Carbonatis
>   Sodae Phosphatis
>   Aquae destillatae
>   quantum sufficit
>   Caloris 150° Centigrade
>   Vacui perfectissimi
>   Patientiae.

Transubstantiation will be nothing to this if it turns out to be true, and you may go and tell your neighbour Januarius to shut up his shop as the heretics mean to out-bid him.

Now I think that the best service I can render to all you enterprising young men is to turn devil's advocate, and do my best to pick holes in your work.

By the way, Miklucho-Maclay [1] has been here ; I have

---

[1] Miklucho-Maclay, a Russian naturalist, and close friend of Haeckel's, who later adventured himself alone among the cannibals of New Guinea.

seen a good deal of him, and he strikes me as a man of very considerable capacity and energy. He was to return to Jena to-day.

My friend Herbert Spencer will be glad to learn that you appreciate his book. I have been *his* devil's advocate for a number of years, and there is no telling how many brilliant speculations I have been the means of choking in an embryonic state.

My wife does not know that I am writing to you, or she would say apropos of your last paragraph that you are an entirely unreasonable creature in your notions of how friendship should be manifested, and that you make no allowances for the oppression and exhaustion of the work entailed by what Jean Paul calls a " Töchtervolles Haus." I hope I may live to see you with at least ten children, and then my wife and I will be avenged. Our children will be married and settled by that time, and we shall have time to write every day and get very wroth when you do not reply immediately. — Ever yours faithfully,   T. H. HUXLEY.

All are well, the children so grown you will not know them.

July 18, 1870.

MY DEAR DOHRN — Notwithstanding the severe symptoms of " Töchterkrankheit" under which I labour, I find myself equal to reply to your letter.

The British Association meets in September on the 14th day of that month, which falls on a Wednesday. Of course, if you come you shall be provided for by the best specimen of Liverpool hospitality. We have ample provision for the entertainment of the "distinguished foreigner."

Will you be so good as to be my special ambassador with Haeckel and Gegenbaur, and tell them the same thing ? It would give me and all of us particular pleasure to see them and to take care of them.

But I am afraid that this wretched war will play the very deuce with our foreign friends. If you Germans do not give that crowned swindler, whose fall I have been looking for ever since the *coup d'état*, such a blow as he will never recover from, I will never forgive you. Public opinion in England is not worth much, but at present, it is entirely against France. Even the *Times*, which general[ly] contrives to be on the baser side of a controversy, is at present on the German side. And my daughters announced to me yesterday that they had converted a young friend of theirs from the French to the German side, which is one gained for you. All look forward with great pleasure to seeing you in the autumn. —Ever yours faithfully, T. H. HUXLEY.

In addition to this address on September 14, he read his paper on "Penicillium," etc., in Section D on the 20th. Speaking on the 17th, after a lecture of Sir J. Lubbock's on the "Social and Religious Condition of the Lower Races of Mankind," he brought forward his own experiences as to the practical results of the beliefs held by the Australian savages, and from this passed to the increasing savagery of the lower classes in great towns such as Liverpool, which was the great political question of the future, and for which the only cure lay in a proper system of education.

The savagery underlying modern civilisation was all the more vividly before him, because one evening he, together with Sir J. Lubbock, Dr. Bastian, and Mr. Samuelson, were taken by the chief of the detective department round some of the worst slums in Liverpool. In thieves' dens, doss houses, dancing saloons, enough of suffering and criminality was seen to leave a very deep and painful impression. In one of these places, a thieves' lodging-house, a drunken man with a cut face accosted him and asked him whether he was a doctor. He said "yes," whereupon the man asked him to doctor his face. He had been fighting, and was terribly excited. Huxley tried to pacify him, but if it had not been for the intervention of the detective, the man would have assaulted him. Afterwards he asked the detective if he were not afraid to go alone in these places, and got the significant answer, "Lord bless you, sir, drink and disease take all the strength out of them."

On the 21st, after the general meeting of the Association, which wound up the proceedings, the Historical Society of Lancashire and Cheshire presented a diploma of honorary membership and a gift of books to Huxley, Sir G. Stokes, and Sir J. Hooker, the last three Presidents of the British Association, and to Professors Tyndall and Rankine and Sir J. Lubbock, the lecturers at Liverpool. Then Huxley was presented with a mazer bowl lined with silver, made from part of one of the roof timbers of the cottage occupied as his headquarters by Prince Rupert during the siege of Liverpool. He was rather taken aback when he found the bowl was filled with champagne; after a moment, however, he drank "success to the good old town of Liverpool," and with a wave of his hand, threw the rest on the floor, saying, "I pour this as a libation to the tutelary deities of the town."

The same evening he was the guest of the Sphinx Club at dinner at the Royal Hotel, his friend Mr. P. H. Rathbone being in the chair, and in proposing the toast of the town and trade of Liverpool, declared that commerce was a greater civiliser than all the religion and all the science ever put together in the world, for it taught men to be truthful and punctual and precise in the execution of their engagements, and men who were truthful and punctual and precise in the execution of their engagements had put their feet upon the first rung of the ladder which led to moral and intellectual elevation.

There were the usual clerical attacks on the address, among the rest a particularly violent one from a Unitarian pulpit. Writing to Mr. Samuelson on October 5 he says :—

Be not vexed on account of the godly. They will have their way. I found Mr. ——'s sermon awaiting me on my return home. It is an able paper, but like the rest of his cloth he will not take the trouble to make himself acquainted with the ideas of the man whom he opposes. At least that is the case if he imagines he brings me under the range of his guns.

On October 2 he writes to Tyndall :—

I have not yet thanked you properly for your great contribution to the success of our meeting [i.e. his lecture "On the Scientific Uses of the Imagination"]. I was nervous over the passage about the clergy, but those confounded parsons seem to me to let you say anything, while they bully me for a word or a phrase. It's the old story, "one man may steal a horse while the other may not look over the wall."

Tyndall was not to be outdone, and replied :—

The parsons know very well that I mean kindness; if I correct them I do it in love and not in wrath.

One more extract from a letter to Dr. Dohrn, under date of November 17. The first part is taken up with a long and detailed description of the best English microscopes and their price, for Dr. Dohrn wished to get one ; and my father volunteered to procure it for him. The rest of the letter has a more general interest as giving his views on the great struggle between France and Germany then in progress, his distrust of militarism, and above all, his hatred of lying, political as much as any other :—

This wretched war is doing infinite mischief, but I do not see what Germany can do now but carry it out to the end.

I began to have some sympathy with the French after Sedan, but the Republic lies harder than the Empire did, and the whole country seems to me to be rotten to the core. The only figure which stands out with anything like nobility or dignity, on the French side, is that of the Empress, and she is only a second-rate Marie-Antoinette. There is no Roland, no Corday, and apparently no *man* of any description.

The Russian row is beginning, and the rottenness of English administration will soon, I suppose, have an opportunity of displaying itself. Bad days are, I am afraid, in store for all of us, and the worst for Germany if it once becomes thoroughly bitten by the military mad dog.

The "happy family" is flourishing and was afflicted, even over its breakfast, when I gave out the news that you had been ill.

The wife desires her best remembrances, and we all hope you are better.

The high pressure under which Huxley worked, and his abundant output, continued undiminished through the autumn and winter. Indeed, he was so busy that he postponed his Lectures to Working Men in London from October to February 1871. On October 3 he lectured in Leicester on "What is to be Learned from a Piece of Coal," a parallel lecture to that of 1868 on "A Piece of Chalk." On the 17th and 24th he lectured at Birmingham on "Extinct Animals intermediate between Reptiles and Birds "— a subject which he had made peculiarly his own by long study ; and on December 29 he was at Bradford, and lectured at the Philosophical Institute upon " The Formation of Coal " (*Coll. Ess.* viii.).

He was also busy with two Royal Commissions ; still, at whatever cost of the energy and time due to his own investigations and those additional labours by which he increased his none too abundant income, he felt it his duty, in the interests of his ideal of education, to come forward as a candidate for the newly-instituted School Board for London. This was the practical outcome of the rising interest in education all over the country; on its working, he felt, depended momentous issues—the fostering of the moral and physical well-being of the nation; the quickening of its intelligence and the maintenance of its commercial supremacy. Withal, he desired to temper "book-learning" with something of the direct knowledge of nature: on the one hand, as an admirable instrument of education, if properly applied; on the other, as preparing the way for an attitude of mind which could appreciate the reasons for the immense changes already beginning to operate in human thought.

Moreover, he possessed a considerable knowledge of the working of elementary education throughout the country, owing to his experience as examiner under the Science and Art Department, the establishment of which he describes as "a measure which came into existence unnoticed, but which will, I believe, turn out to be of more importance to the welfare of the people than many political changes over which the noise of battle has rent the air" (*Scientific Education*, 1869; *Coll. Ess.* iii. p. 131).

Accordingly, though with health uncertain, and in the midst of exacting occupations, he felt that he ought not to stand aside at so critical a moment, and offered himself for election in the Marylebone division with a secret sense that rejection would in many ways be a great relief.

The election took place on November 29, and Huxley came out second on the poll. He had had neither the means nor the time for a regular canvass of the electors. He was content to address several public meetings, and leave the result to the interest he could awaken amongst his hearers. His views were further brought before the public by the action of the editor of the *Contemporary Review*, who, before the election, "took upon himself, in what seemed to him to be the public interest," to send to the newspapers an extract from Huxley's article, "The School Boards: what they can do, and what they may do," which was to appear in the December number.

In this article will be found (*Coll. Ess.* iii. p. 374) a full account of the programme which he laid down for himself, and which to a great extent he saw carried into effect, in its fourfold division—of physical drill and discipline, not only to improve the physique of the children, but as an introduction to all other sorts of training—of domestic training, especially for girls —of education in the knowledge of moral and social laws and the engagement of the affections for what is good and against what is evil—and finally, of intellectual training. And it should be noted that he did not only regard intellectual training from the utilitarian point of view; he insisted, *e.g.* on the value of reading for amusement as "one of its most valuable uses to hard-worked people."

Much as he desired that this intellectual training should be efficient, the most cursory perusal of this article will show how far he placed the moral training above the intellectual, which, by itself, would only turn the gutter-child into "the subtlest of all the beasts of the field," and how wide of the mark is the cartoon at this period representing him as the professor whose panacea for the ragged children was to " cram them full of nonsense."

In the third section are also to be found his arguments for the retention of Bible-reading in the elementary schools. He reproached extremists of either party for confounding the science, theology, with the affection, religion, and either crying for more theology under the name of religion, or demanding the abolition of "religious" teaching in order to get rid of theology, a step which he likens to "burning your ship to get rid of the cockroaches."

As regards his actual work on the Board, I must express my thanks to Dr. J. H. Gladstone for his kindness in supplementing my information with an account based partly on his own long experience of the Board, partly on the reminiscences of members contemporary with my father.

The Board met first on December 15, for the purpose of electing a Chairman. As a preliminary, Huxley proposed and carried a motion that no salary be attached to the post. He was himself one of the four members proposed for the Chairmanship; but the choice of the Board fell upon Lord Lawrence. In the words of Dr. Gladstone :—

Huxley at once took a prominent part in the proceedings, and continued to do so till the beginning of the year 1872, when ill-health compelled him to retire.

At first there was much curiosity both inside and outside the Board as to how Huxley would work with the old educationists, the clergy, dissenting ministers, and the miscellaneous body of eminent men that comprised the first Board. His antagonism to many of the methods employed in elementary schools was well known from his various discourses, which had been recently published together under the title of *Lay Sermons, Addresses, and Reviews.* I watched his course with interest at the time; but for the purpose of this sketch I have lately sought information from such of the old members of the Board as are still living, especially the Earl of Harrowby, Bishop Barry, the Rev. Dr. Angus, and Mr. Edward North Buxton, together with Mr. Croad, the Clerk of the Board. They soon found proof of his great energy, and his power of expressing his views in clear and forcible language; but they also found that with all his strong convictions and lofty ideals he was able and willing to enter into the views of others, and to look at a practical question from its several sides. He could construct as well as criticise. Having entered a public arena somewhat late in life, and being of a sensitive nature, he had scarcely acquired that calmness and pachydermatous quality which is needful for one's personal comfort; but his colleagues soon came to respect him as a perfectly honest antagonist or supporter, and one who did not allow differences of conviction to interfere with friendly intercourse.

The various sections of the clerical party indeed looked forward with great apprehension to his presence on the Board, but the more liberal amongst them ventured to find ground for hoping that they and he would not be utterly opposed so far as the work of practical organisation was concerned, in the declaration of his belief that true education was impossible without "religion," of which he declared

that all that has an unchangeable reality in it is
constituted by the love of some ethical ideal to
govern and guide conduct, "together with the awe
and reverence, which have no kinship with base fear,
but rise whenever one tries to pierce below the
surface of things, whether they be material or
spiritual." And in fact a cleavage took place between
him and the seven extreme "secularists" on the
Board (the seven champions of unchristendom, as
their opponents dubbed them) on the question of the
reading of the Bible in schools (see below, p.
31).[1]

One of the earliest proposals laid before the Board
was a resolution to open the meetings with prayer.
To this considerable opposition was offered; but a
bitter debate was averted by Huxley pointing out
that the proposal was *ultra vires*, inasmuch as under
the Act constituting the Board the business for which
they were empowered to meet did not include prayer.
Hereupon a requisition—in which he himself joined—
was made to allow the use of a committee-room to
those who wished to unite in a short service before
the weekly meetings, an arrangement which has
continued to the present time.

At the second meeting, on December 21, he gave
notice of a motion to appoint a committee to consider

---

[1] Bishop Barry calls particular attention to his attitude on this
point, "because," he says, "it is (I think) often misunderstood.
In the *Life* (for instance) *of the Right Honourable W. H. Smith*,
published not long ago, Huxley is supposed, as a matter of course,
to have been the leader of the Secularist party."

and report upon the scheme of education to be
adopted in the Board Schools.

This motion came up for consideration on February
15, 1871. In introducing it, he said that such a
committee ought to consider—

First, the general nature and relations of the schools
which may come under the Board. Secondly, the amount
of time to be devoted to educational purposes in such
schools; and Thirdly, the subject-matter of the instruction or education, or teaching, or training, which is to be
given in these schools.

But this, by itself, he continued, would be in-
complete. At one end of the scale he advocated
Infant schools, and urged a connection with the
excellent work of the Ragged schools. At the other
end he desired to see continuation schools, and
ultimately some scheme of technical education. A
comprehensive scheme, indeed, would involve an
educational ladder from the gutter to the university,
whereby children of exceptional ability might reach
the place for which nature had fitted them.

The subject matter of elementary instruction must
be limited by what was practicable and desirable.
The revised code had done too little; it had taught
the use of the tools of learning, while denying all
sorts of knowledge on which to exercise them afterwards. And here incidentally he repudiated the
notion that the English child was stupid; on the
contrary, he thought the two finest intellects in
Europe at this time were the English and the Italian.

In particular he advocated the teaching of "the first elements of physical science"; "by which I do not mean teaching astronomy and the use of the globes, and the rest of the abominable trash—but a little instruction of the child in what is the nature of common things about him; what their properties are, and in what relation this actual body of man stands to the universe outside of it." "There is no form of knowledge or instruction in which children take greater interest."

Drawing and music, too, he considered, should be taught in every elementary school, not to produce painters or musicians, but as civilising arts.   History, except the most elementary notions, he put out of court, as too advanced for children

Finally, he proposed a list of members to serve on the Education Committee in a couple of sentences with a humorous twist in them which disarmed criticism. "On a former occasion I was accused of having a proclivity in favour of the clergy, and recollecting this, I have only given them in this instance a fair proportion of the representation. If, however, I have omitted any gentleman who thinks he ought to be on the committee, I can only assure him that above all others I should have been glad to put him on."

That day week the committee was elected, about a third of the members of the Board being chosen to serve on it.   At the same meeting, Dr. Gladstone continues—

Mr. W. H. Smith, the well-known member of
Parliament, proposed, and Mr. Samuel Morley, M.P.,
seconded, a resolution in favour of religious teaching—
" That, in the schools provided by the Board, the Bible
shall be read, and there shall be given therefrom such
explanations and such instruction in the principles of
religion and morality as are suited to the capacities of
children," with certain provisos.   Several antagonistic
amendments were proposed ; but Prof. Huxley gave his
support to Mr. Smith's resolutions, which, however, he
thought might be trimmed and amended in a way that
the Rev. Dr. Angus had suggested.   His speech, defining
his own position, was a very remarkable one.   He said " it
was assumed in the public mind that this question of
religious instruction was a little family quarrel between
the different sects of Protestantism on the one hand, and
the old Catholic Church on the other.   Side by side with
this much shivered and splintered Protestantism of theirs,
and with the united fabric of the Catholic Church (not
so strong temporally as she used to be, otherwise he might
not have been addressing them at that moment), there
was a third party growing up into very considerable and
daily increasing significance, which had nothing to do
with either of those great parties, and which was pushing
its own way independent of them, having its own religion
and its own morality, which rested in no way whatever
on the foundations of the other two."   He thought that
" the action of the Board should be guided and influenced
very much by the consideration of this third great aspect
of things," which he called the scientific aspect, for want
of a better name.

" It had been very justly said that they had a great
mass of low half-instructed population which owed what
little redemption from ignorance and barbarism it
possessed mainly to the efforts of the clergy of the
different denominations.   Any system of gaining the
attention of these people to these matters must be a

system connected with, or not too rudely divorced from their own system of belief. He wanted regulations, not in accordance with what he himself thought was right, but in the direction in which thought was moving." He wanted an elastic system, that did not oppose any obstacle to the free play of the public mind.

Huxley voted against all the proposed amendments, and in favour of Mr. Smith's motion. There were only three who voted against it; while the three Roman Catholic members refrained from voting. This basis of religious instruction, practically unaltered, has remained the law of the Board ever since.

There was a controversy in the papers, between Prof. Huxley and the Rev. W. H. Fremantle, as to the nature of the explanations of the Bible lessons. Huxley maintained that it should be purely grammatical, geographical, and historical in its nature; Fremantle that it should include some species of distinct religious teaching, but not of a denominational character.[1]

In taking up this position, Huxley expressly disclaimed any desire for a mere compromise to smooth over a difficulty. He supported what appeared to be the only workable plan under the circumstances, though it was not his ideal; for he would not have used the Bible as the agency for introducing the religious and ethical idea into education if he had been dealing with a fresh and untouched population.

His appreciation of the literary and historical value of the Bible, and the effect it was likely to

[1] Cp. extract from Lord Shaftesbury's journal about this correspondence (*Life and Work of Lord Shaftesbury*, iii. 282). "Professor Huxley has this definition of morality and religion: 'Teach a child what is wise, that is *morality*. Teach him what is wise and beautiful, that is *religion!*' Let no one henceforth despair of making things clear and of giving explanations !"

produce upon the school children, circumstanced as they were, is sometimes misunderstood to be an endorsement of the vulgar idea of it. But it always remained his belief "that the principle of strict secularity in State education is sound, and must eventually prevail." [1]

His views on dogmatic teaching in State schools, may be gathered further from two letters at the period when an attempt was being made to upset the so-called compromise.

The first appeared in the *Times* of April 29, 1893 :—

Sir—In a leading article of your issue of to-day you state, with perfect accuracy, that I supported the arrangement respecting religious instruction agreed to by the London School Board in 1871, and hitherto undisturbed. But you go on to say that "the persons who framed the rule" intended it to include definite teaching of such theological dogmas as the Incarnation.

I cannot say what may have been in the minds of the framers of the rule ; but, assuredly, if I had dreamed that any such interpretation could fairly be put upon it, I should have opposed the arrangement to the best of my ability.

In fact, a year before the rule was framed I wrote an article in the *Contemporary Review*, entitled "The School

---

[1] As a result of some remarks of Mr. Clodd's on the matter in *Pioneers of Evolution*, a correspondent, some time after, wrote to him as follows :—

"In the report upon State Education in New Zealand, 1895, drawn up by R. Laishly, the following occurs, p. 13 :— 'Professor Huxley gives me leave to state his opinion to be that the principle of strict secularity in State education is sound, and must eventually prevail.'"

Boards—what they can do and what they may do," in which I argued that the terms of the Education Act excluded such teaching as it is now proposed to include. And I support my contention by the following citation from a speech delivered by Mr. Forster at the Birkbeck Institution in 1870 :—

> I have the fullest confidence that in the reading and explaining of the Bible what the children will be taught will be the great truths of Christian life and conduct, which all of us desire they should know, and that no efforts will be made to cram into their poor little minds theological dogmas which their tender age prevents them from understanding.—I am, Sir, your obedient servant,
>
> T. H. HUXLEY.

HODESLEA, EASTBOURNE, *April* 28.

The second is to a correspondent who wrote to ask him whether adhesion to the compromise had not rendered nonsensical the teaching given in a certain lesson upon the finding of the youthful Jesus in the temple, when, after they had read the verse, "How is it that ye sought me?   Wist ye not that I must be about my Father's business?" the teacher asked the children the name of Jesus' father and mother, and accepted the simple answer, Joseph and Mary.   Thus the point of the story, whether regarded as reality or myth, is slurred over, the result is perplexity, the teaching, in short, is bad, apart from all theory as to the value of the Bible.

In a letter to the *Chronicle*, which he forwarded, this correspondent suggested a continuation of the " incriminated lesson " :—

Suppose, then, that an intelligent child of seven, who has just heard it read out that Jesus excused Himself to His parents for disappearing for three days, on the ground that He was about His Father's business, and has then learned that His father's name was Joseph, had said, " Please, teacher, was this the Jesus that gave us the Lord's Prayer ? " The teacher answers, " Yes." And suppose the child rejoins, "And is it to His father Joseph that he bids us pray when we say Our Father ? " But there are boys of nine, ten, eleven years in Board Schools, and many such boys are intelligent enough to take up the subject of the lesson where the instructor left it. " Please, teacher," asks one of these, " what business was it that Jesus had to do for His father Joseph ? Had He stopped behind to get a few orders ? Was it true that He had been about Joseph's business ? And, if it was not true, did He not deserve to be punished ? "

Huxley replied on October 16, 1894 :—

DEAR SIR—I am one with you in hating "hush up " as I do all other forms of lying ; but I venture to submit that the compromise of 1871 was not a "hush up." If I had taken it to be such I should have refused to have anything to do with it. And more specifically, I said in a letter to the *Times* (see *Times*, 29th April 1893) at the beginning of the present controversy, that if I had thought the compromise involved the obligatory teaching of such dogmas as the Incarnation I should have opposed it.

There has never been the slightest ambiguity about my position in this matter ; in fact, if you will turn to one paper on the School Board written by me before my election in 1870, I think you will find that I anticipated the pith of the present discussion.

The persons who agreed to the compromise, did exactly what all sincere men who agree to compromise, do. For the sake of the enormous advantage of giving the rudiments of a decent education to several generations of the people, they accepted what was practically an armistice in respect of certain matters about which the contending parties were absolutely irreconcilable.

The clericals have now "denounced" the treaty, doubtless thinking they can get a new one more favourable to themselves.

From my point of view, I am not sure that it might not be well for them to succeed, so that the sweep into space which would befall them in the course of the next twenty-three years might be complete and final.

As to the case you put to me—permit me to continue the dialogue in another shape.

Boy.—Please, teacher, if Joseph was not Jesus' father and God was, why did Mary say, "Thy father and I have sought thee sorrowing"? How could God not know where Jesus was? How could He be sorry?

Teacher.—When Jesus says Father, he means God; but when Mary says father, she means Joseph.

Boy.—Then Mary didn't know God was Jesus' father?

Teacher.—Oh yes, she did (reads the story of the Annunciation).

Boy.—It seems to me very odd that Mary used language which she knew was not true, and taught her son to call Joseph father. But there's another odd thing about her. If she knew her child was God's son, why was she alarmed about his safety. Surely she might have trusted God to look after his own son in a crowd.

I know of children of six and seven who are quite capable of following out such a line of inquiry with all the severe logic of a moral sense which has not been sophisticated by pious scrubbing.

I could tell you of stranger inquiries than these which have been made by children in endeavouring to understand the account of the miraculous conception.

Whence I conclude that even in the interests of what people are pleased to call Christianity (though it is my firm conviction that Jesus would have repudiated the doctrine of the Incarnation as warmly as that of the Trinity), it may be well to leave things as they are.

All this is for your own eye. There is nothing in substance that I have not said publicly, but I do not feel called upon to say it over again, or get mixed up in an utterly wearisome controversy.—I am, yours faithfully,

T. H. HUXLEY.

However, he was unsuccessful in his proposal that a selection be made of passages for reading from the Bible ; the Board refused to become censors. On May 10 he raised the question of the diversion from the education of poor children of charitable bequests, which ought to be applied to the augmentation of the school fund. In speaking to this motion he said that the long account of errors and crimes of the Catholic Church was greatly redeemed by the fact that that Church had always borne in mind the education of the poor, and had carried out the great democratic idea that the soul of every man was of the same value in the eyes of his Maker.

The next matter of importance in which he took part was on June 14, when the Committee on the Scheme of Education presented its first report. Dr. Gladstone writes :—

It was a very voluminous document. The Committee had met every week, and, in the words of Huxley, " what it had endeavoured to do, was to obtain some order and system and uniformity in important matters, whilst in comparatively unimportant matters they thought some play should be given for the activity of the bodies of men into whose hands the management of the various schools should be placed." The recommendations were considered on June 21 and July 12, and passed without any material alterations or additions. They were very much the same as existed in the best elementary schools of the period. Huxley's chief interest, it may be surmised, was in the subjects of instruction. It was passed that, in infants' schools there should be the Bible, reading, writing, arithmetic, object lessons of a simple character, with some such exercise of the hands and eyes as is given in the Kindergarten system, music, and drill. In junior and senior schools the subjects of instruction were divided into two classes, essential and discretionary, the essentials being the Bible, and the principles of religion and morality, reading, writing, and arithmetic, English grammar and composition, elementary geography, and elementary social economy, history of England, the principles of book-keeping in senior schools, with mensuration in senior boys' schools. All through the six years there were to be systematised object lessons, embracing a course of elementary instruction in physical science, and serving as an introduction to the science examinations conducted by the Science and Art department. An analogous course of instruction was adopted for elementary evening schools. In moving "that the formation of science and art classes in connection with public elementary schools be encouraged and facilitated," Huxley contended strongly for it, saying, "The country could not possibly commit a greater error than in establishing schools in which the direct applications of science and art were taught before those who entered the classes were grounded in the principles of physical science." In advocating object lessons he said, "The position that science was now assuming, not only in relation to practical life, but to thought, was such that those who remained entirely ignorant of even its elementary facts were in a wholly unfair position as regarded the world of thought and the world of practical life." It was, moreover, "the only real foundation for technical education."

Other points in which he was specially concerned were, that the universal teaching of drawing was accepted, against an amendment excluding girls; that domestic economy was made a discretionary substitute for needlework and cutting-out; while he spoke in defence of Latin as a discretionary subject, alternatively with a modern language.   It was true that he would not have proposed it in the first instance, not because a little Latin is a bad thing, but for fear of "overloading the boat."   But, on the other hand, there was great danger if education were not thrown open to all without restriction.   If it be urged that a man should be content with the state of life to which he is called, the obvious retort is, How do you know what is your state of life, unless you try what you are called to? There is no more frightful "sitting on the safety valve" than in preventing men of ability from having the means of rising to the positions for which they, by their talents and industry, could qualify themselves.

Further, although the committee as a whole recommended that discretionary subjects should be extras, he wished them to be covered by the general payment, in which sense the report was amended.

This Education Committee (proceeds Dr. Gladstone) continued to sit, and on November 30 brought up a report in favour of the Prussian system of separate classrooms, to be tried in one school as an experiment.   This reads curiously now that it has become the system almost universally adopted in the London Board Schools.

In regard to examinations Huxley strongly supported

the view that the teaching in all subjects, secular or sacred, should be periodically tested.

On December 13, Huxley raised the question whether the selection of books and apparatus should be referred to his Committee or to the School Management Committee, and on January 10 following, a small sub-committee for that object was formed. Almost immediately after this he retired from the Board.

One more speech of his, which created a great stir at the time, must be referred to, namely his expression of undisguised hostility to the system of education maintained by the Ultramontane section of the Roman Catholics.[1] In October the bye-laws came up for consideration. One of them provided that the Board should pay over direct to denominational schools the fees for poor children. This he opposed on the ground that it would lead to repeated contests on the Board, and further, might be used as a tool by the Ultramontanes for their own purposes. Believing that their system as set forth in the syllabus, of securing complete possession of the minds of those whom they taught or controlled, was destructive to all that was highest in the nature of mankind, and inconsistent with intellectual and political liberty, he considered it his earnest duty to oppose all measures which would lead to assisting the Ultramontanes in their purpose.

Hereupon he was vehemently attacked, for example, in the *Times* for his "injudicious and even reprehensible tone" which "aggravated the difficulties his

---

[1] Cp. "Scientific Education," *Coll. Ess.* iii. p. 111.

opponents might have in giving way to him." Was this, it was asked, the way to get Roman Catholic children to the Board schools? Was it not an abandonment of the ideal of compulsory education?

It is hardly necessary to point out that the question was not between the compulsory inclusion or exclusion of poor children, but between their admission at the cost of the Board to schools under the Board's own control or outside it. In any case the children of Roman Catholics were not likely to get their own doctrines taught in Board Schools, and without this they declared they would rather go without education at all.

Early in 1872 Huxley retired. For a year he had continued at this task; then his health broke down, and feeling that he had done his part, from no personal motives of ambition, but rather at some cost to himself, for what he held to be national ends, he determined not to resume the work after the rest which was to restore him to health, and made his resignation definite.

Dr. Gladstone writes :——

On February 7 a letter of resignation was received from him, stating that he was "reluctantly compelled, both on account of his health and his private affairs, to insist on giving up his seat at the Board." The Rev. Dr. Rigg, Canon Miller, Mr. Charles Reed, and Lord Lawrence expressed their deep regret. In the words of Dr. Rigg, "they were losing one of the most valuable members of the Board, not only because of his intellect and trained acuteness, but because of his knowledge of every subject connected with culture and education, and because of his great fairness and impartiality with regard to all subjects that came under his observation."

Though Huxley quitted the Board after only fourteen months' service, the memory of his words and acts combined to influence it long afterwards. In various ways he expressed his opinion on educational matters, publicly and privately. He frequently talked with me on the subject at the Athenæum Club, and shortly after my election to the Board in 1873, I find it recorded in my diary that he insisted strongly on the necessity of our building infants' schools,—"people may talk about intellectual teaching, but what we principally want is the moral teaching."

As to the sub-committee on books and apparatus, it did little at first, but at the beginning of the second Board, 1873, it became better organised under the presidency of the Rev. Benjamin Waugh. At the commencement of the next triennial term I became the chairman, and continued to be such for eighteen years. It was our duty to put into practice the scheme of instruction which Huxley was mainly instrumental in settling. We were thus able indirectly to improve both the means and methods of teaching. The subjects of instruction have all been retained in the Curriculum of the London School Board, except, perhaps, "mensuration" and "social economy." The most important developments and additions have been in the direction of educating the hand and eye. Kindergarten methods have been promoted. Drawing, on which Huxley laid more stress than his colleagues generally did, has been enormously extended and greatly revolutionised in its methods. Object lessons and elementary science have been introduced everywhere, while shorthand, the use of tools for boys, and cookery and domestic economy for girls are becoming essentials in our schools. Evening continuation schools have lately been widely extended. Thus the impulse given by

Huxley in the first months of the Board's existence has been carried forward by others, and is now affecting the minds of the half million of boys and girls in the Board Schools of London, and indirectly the still greater number in other schools throughout the land.

I must further express my thanks to Bishop Barry for permission to make use of the following passages from the notes contributed by him to Dr. Gladstone:—

I had the privilege of being a member of his committee for defining the curriculum of study, and here also—the religious question being disposed of—I was able to follow much the same line as his, and I remember being struck not only with his clear-headed ability, but with his strong commonsense, as to what was useful and practicable, and the utter absence in him of *doctrinaire* aspiration after ideal impossibilities. There was (I think) very little under his chairmanship of strongly-accentuated difference of opinion.

In his action on the Board generally I was struck with these three characteristics :—First, his remarkable power of speaking—I may say, of oratory—not only on his own scientific subjects, but on all the matters, many of which were of great practical interest and touched the deepest feelings, which came before the Board at that critical time. Had he chosen—and we heard at that time that he was considering whether he should choose —to enter political life, it would certainly have made him a great power, possibly a leader, in that sphere. Next, what constantly appears in his writings, even those of the most polemical kind—a singular candour in recognising truths which might seem to militate against his own position, and a power of understanding and respecting his adversaries' opinions, if only they were strongly and conscientiously held. I remember his saying on one occasion that in his earlier experience of sick-

ness and suffering, he had found that the most effective
helpers of the higher humanity were not the scientist or
the philosopher, but "the parson, and the sister, and the
Bible woman." Lastly, the strong commonsense, which
enabled him to see what was "within the range of
practical politics," and to choose for the cause which he
had at heart the line of least resistance, and to check,
sometimes to rebuke, intolerant obstinacy even on the
side which he was himself inclined to favour. These
qualities over and above his high intellectual ability
made him, for the comparatively short time that he
remained on the Board, one of its leading members.

No less vivid is the impression left, after many
years, upon another member of the first School
Board, the Rev. Benjamin Waugh, whose life-long
work for the children is so well known. From
his recollections, written for the use of Professor
Gladstone, it is my privilege to quote the following
paragraphs:—

I was drawn to him most, and was influenced by him
most, because of his attitude to a child. He was on the
Board to establish schools for children. His motive in
every argument, in all the fun and ridicule he indulged
in, and in his occasional anger, was the child. He
resented the idea that schools were to train either
congregations for churches or hands for factories. He
was on the Board as a friend of children. What he
sought to do for the child was for the child's sake, that
it might live a fuller, truer, worthier life. If ever his
great tolerance with men with whom he differed on
general principles seemed to fail him for a moment, it
was because they seemed to him to seek other ends than
the child for its own sake. . . .
His contempt for the idea of the world into which

we were born being either a sort of clergyhouse or a
market-place, was too complete to be marked by any
eagerness. But in view of the market-place idea he was
the less calm.

Like many others who had not yet come to know in
what high esteem he held the moral and spiritual nature
of children, I had thought he was the advocate of mere
secular studies, alike in the nation's schools, and in its
families. But by contact with him, this soon became
an impossible idea. In very early days on the Board a
remark I had made to a mutual friend which implied this
unjust idea was repeated to him. "Tell Waugh that he
talks too fast," was his message to me. I was not long
in finding out that this was a very just reproof. . . .

The two things in his character of which I became
most conscious by contact with him, were his childlike-
ness and his consideration for intellectual inferiors.
His arguments were as transparently honest as the
arguments of a child. They might or might not seem
wrong to others, but they were never untrue to himself.
Whether you agreed with them or not, they always
added greatly to the charm of his personality. Whether
his face was lighted by his careless and playful
humour or his great brows were shadowed by anger, he
was alike expressing himself with the honesty of a child.
What he counted iniquity he hated, and what he counted
righteous he loved with the candour of a child. . . .

Of his consideration for intellectual inferiors I, of
course, needed a large share, and it was never wanting.
Towering as was his intellectual strength and keenness
above me, indeed above the whole of the rest of the
members of the Board, he did not condescend to me.
The result was never humiliating. It had no pain of
any sort in it. He was too spontaneous and liberal with
his consideration to seem conscious that he was showing
any. There were many men of religious note upon the
Board, of some of whom I could not say the same.

In his most trenchant attacks on what he deemed
wrong in principles, he never descended to attack either
the sects which held them or the individuals who
supported them, even though occasionally much provocation was given him. He might not care for peace with
some of the theories represented on the Board, but he
had certainly and at all times great good-will to men.

As a speaker he was delightful. Few, clear, definite,
and calm as stars were the words he spoke. Nobody
talked whilst he was speaking. There were no tricks
in his talk. He did not seem to be trying to persuade
you of something. What convinced him, that he
transferred to others. He made no attempt to mis-
represent those opposed to him. He sought only to let
them know himself. . . . Even the sparkle of his
humour, like the sparkle of a diamond, was of the
inevitable in him, and was as fair as it was enjoyable.

As one who has tried to serve children, I look back
upon having fallen in with Mr. Huxley as one of the
many fortunate circumstances of my life. It taught me
the importance of making acquaintance with facts, and
of studying the laws of them. Under his influence it was
that I most of all came to see the practical value of a single
eye to those in any pursuit of life. I saw what effect
they had on emotions of charity and sentiments of justice,
and what simplicity and grandeur they gave to appeals.

My last conversation with him was at Eastbourne
some time in 1887 or 1888. I was there on my society's
business. "Well, Waugh, you're still busy about your
babies," was his greeting. "Yes," I responded, "and you
are still busy about your pigs." One of the last discussions at which he was present at the School Board for
London had been on the proximity of a piggery to a
site for a school, and his attack on Mr. Gladstone on the
Gadarene swine had just been made in the *Nineteenth
Century.* "Do you still believe in Gladstone?" he continued. "That man has the greatest intellect in Europe.

He was born to be a leader of men, and he has debased
himself to be a follower of the masses. If working men
were to-day to vote by a majority that two and two made
five, to-morrow Gladstone would believe it, and find them
reasons for it which they had never dreamed of." He
said it slowly and with sorrow.

Two more incidents are connected with his service
on the School Board. A wealthy friend wrote to
him in the most honourable and delicate terms,
begging him, on public grounds, to accept £400 a
year to enable him to continue his work on the
Board. He refused the offer as simply and straight-
forwardly as it was made; his means, though not
large, were sufficient for his present needs.

Further, a good many people seemed to think that
he meant to use the School Board as a stalking horse
for a political career. To one of those who urged
him to stand for Parliament, he replied thus :—

Nov. 18, 1871.

DEAR SIR—It has often been suggested to me that I
should seek for a seat in the House of Commons ; indeed
I have reason to think that many persons suppose that I
entered the London School Board simply as a road to
Parliament.

But I assure you that this supposition is entirely
without foundation, and that I have never seriously
entertained any notion of the kind.

The work of the School Board involves me in no
small sacrifices of various kinds, but I went into it with
my eyes open, and with the clear conviction that it was
worth while to make those sacrifices for the sake of helping the Education Act into practical operation. A year's
experience has not altered that conviction ; but now that

the most difficult, if not the most important, part of our
work is done, I begin to look forward with some anxiety
to the time when I shall be relieved of duties which so
seriously interfere with what I regard as my proper
occupation.

No one can say what the future has in store for him,
but at present I know of no inducement, not even the
offer of a seat in the House of Commons, which would
lead me, even temporarily and partially, to forsake that
work again.—I am, dear sir, yours very faithfully,

T. H. HUXLEY.

I give here a letter to me from Sir Mountstuart
Grant Duff, who also at one period was anxious to
induce him to enter Parliament :—

LEXDEN PARK, COLCHESTER,
4th November 1898.

DEAR MR. HUXLEY—I have met men who seemed to
me to possess powers of mind even greater than those of
your father—his friend Henry Smith for example ; but I
never met any one who gave me the impression so much
as he did, that he would have gone to the front in any
pursuit in which he had seen fit to engage. Henry Smith
had, in addition to his astonishing mathematical genius,
and his great talents as a scholar, a rare faculty of
persuasiveness. Your father used to speak with much
admiration and some amusement of the way in which he
managed to get people to take his view by appearing to
take theirs ; but he never could have been a power in
a popular assembly, nor have carried with him by the
force of his eloquence, great masses of men. I do not
think that your father, if he had entered the House of
Commons and thrown himself entirely into political life,
would have been much behind Gladstone as a debater, or
Bright as an orator. Whether he had the *stamina* which
are required not only to reach but to retain a foremost

place in politics, is another question. The admirers of Prince Bismarck would say that the daily prayer of the statesman should be for "une bonne digestion et un mauvais cœur." "Le mauvais cœur" does not appear to be "de toute necessité," but, assuredly, the "bonne digestion" is. Given an adequate and equal amount of ability in two men who enter the House of Commons together, it is the man of strong digestion, drawing with it, as it usually does, good temper and power of continuous application, who will go furthest. Gladstone, who was inferior to your father in intellect, might have "given points" to the Dragon of Wantley who devoured church steeples. Your father could certainly not have done so, and in that respect was less well equipped for a lifelong parliamentary struggle.

I should like to have seen these two pitted against each other with that "substantial piece of furniture" between them behind which Mr. Disraeli was glad to shelter himself. I should like to have heard them discussing some subject which they both thoroughly understood. When they did cross swords the contest was like nothing that has happened in our times save the struggle at Omdurman. It was not so much a battle as a massacre, for Gladstone had nothing but a bundle of antiquated prejudices wherewith to encounter your father's luminous thought and exact knowledge.

You know, I daresay, that Mr. William Rathbone, then M.P. for Liverpool, once proposed to your father to be the companion of my first Indian journey in 1874-5, he, William Rathbone, paying all your father's expenses.[1] Mr. Rathbone made this proposal when he found that Lubbock, with whom I travelled a great deal at that

---

[1] Of this, Dr. Tyndall wrote to Mrs. Huxley :—"I want to tell you a pleasant conversation I had last night with Jodrell. He and a couple more want to send Hal with Grant Duff to India, taking charge of his duties here and of all necessities ghostly and bodily there !"

period of my life, was unable to go with me to India.
How I wish your father had said "Yes." My journey, as
it was, turned out most instructive and delightful ; but
to have lived five months with a man of his extraordinary
gifts would have been indeed a rare piece of good fortune,
and I should have been able also to have contributed to
the work upon which you are engaged a great many facts
which would have been of interest to your readers.   You
will, however, I am sure, take the will for the deed, and
believe me, very sincerely yours,

M. E. GRANT DUFF.

CHAPTER II

1871

"In 1871" (to quote Sir M. Foster), "the post of Secretary to the Royal Society became vacant through the resignation of William Sharpey, and the Fellows learned with glad surprise that Huxley, whom they looked to rather as a not distant President, was willing to undertake the duties of the office." This office, which he held until 1880, involved him for the next ten years in a quantity of anxious work, not only in the way of correspondence and administration, but the seeing through the press and often revising every biological paper that the Society received, as well as reading those it rejected. Then, too, he had to attend every general, council, and committee meeting, amongst which latter the *Challenger* Committee was a load in itself. Under pressure of all this work, he was compelled to give up active connection with other learned societies.[1]

Other work this year, in addition to the School Board, included courses of lectures at the London

[1] See Appendix II.

Institution in January and February, on "First
Principles of Biology," and from October to December
on "Elementary Physiology"; lectures to Working
Men in London from February to April, as well as
one at Liverpool, March 25, on "The Geographical
Distribution of Animals"; two lectures at the Royal
Institution, May 12 and 19, on "Berkeley on Vision,"
and the " Metaphysics of Sensation" (*Coll. Ess.* vi.).
He published one paleontological paper, "Fossil
Vertebrates from the Yarrow Colliery " (Huxley and
Wright, *Irish Acad. Trans.*).   In June and July he
gave 36 lectures to schoolmasters—that important
business of teaching the teachers that they might set
about scientific instruction in the right way.[1]   He
attended the British Association at Edinburgh, and
laid down his Presidency; he brought out his
"Manual of Vertebrate Anatomy," and wrote a review
of "Mr. Darwin's Critics " (see p. 62 *sq.*), while on
October 9 he delivered an address at the Midland
Institute, Birmingham, on "Administrative Nihilism"
(*Coll. Ess.* i.).   This address, written between September
21 and 28, and remodelled later, was a pendant to
his educational campaign on the School Board ; a
restatement and justification of what he had said and
done there.   His text was the various objections
raised to State interference with education ; he dealt
first with the upholders of a kind of caste system,
men who were willing enough to raise themselves and
their sons to a higher social plane, but objected on

[1] See pp. 59, 80, *sq.*

semi-theological grounds to any one from below doing likewise—neatly satirising them and their notions of gentility, and quoting Plato in support of his contention that what is wanted even more than means to help capacity to rise is " machinery by which to facilitate the descent of incapacity from the higher strata to the lower." He repeats in new phrase his warning " that every man of high natural ability, who is both ignorant and miserable, is as great a danger to society as a rocket without a stick is to people who fire it. Misery is a match that never goes out ; genius, as an explosive power, beats gunpowder hollow : and if knowledge, which should give that power guidance, is wanting, the chances are not small that the rocket will simply run amuck among friends and foes."

Another class of objectors will have it that government should be restricted to police functions, both domestic and foreign, that any further interference must do harm.

Suppose, however, for the sake of argument, that we accept the proposition that the functions of the State may be properly summed up in the one great negative commandment—" Thou shalt not allow any man to interfere with the liberty of any other man,"—I am unable to see that the logical consequence is any such restriction of the power of Government, as its supporters imply. If my next-door neighbour chooses to have his drains in such a state as to create a poisonous atmosphere, which I breathe at the risk of typhoid and diphtheria, he restricts my just freedom to live just as much as if he

went about with a pistol threatening my life ; if he is to
be allowed to let his children go unvaccinated, he might
as well be allowed to leave strychnine lozenges about in
the way of mine ; and if he brings them up untaught
and untrained to earn their living, he is doing his best to
restrict my freedom, by increasing the burden of taxation
for the support of gaols and workhouses, which I have to
pay.

The higher the state ot civilisation, the more completely do the actions of one member of the social body
influence all the rest, and the less possible is it for any
one man to do a wrong thing without interfering, more
or less, with the freedom of all his fellow-citizens.   So
that, even upon the narrowest view of the functions of
the State, it must be admitted to have wider powers than
the advocates of the police theory are disposed to admit.

This leads to a criticism of Mr. Spencer's elaborate
comparison of the body politic to the body physical,
a comparison vitiated by the fact that "among the
higher physiological organisms there is none which is
developed by the conjunction of a number of primi-
tively independent existences into a complete whole."

The process of social organisation appears to be comparable, not so much to the process of organic development, as to the synthesis of the chemist, by which
independent elements are gradually built up into complex
aggregations—in which each element retains an independent individuality, though held in subordination to
the whole.

It is permissible to quote a few more sentences
from this address for the sake of their freshness, or
as illustrating the writer's ideas.

Discussing toleration, "I cannot discover that

Locke fathers the pet doctrine of modern Liberalism, that the toleration of error is a good thing in itself, and to be reckoned among the cardinal virtues."[1]

Of· Mr. Spencer's comparison of the State to a living body in the interests of individualism :—

I suppose it is universally agreed that it would be useless and absurd for the State to attempt to promote friendship and sympathy between man and man directly. But I see no reason why, if it be otherwise expedient, the State may not do something towards that end indirectly. For example, I can conceive the existence of an Established Church which should be a blessing to the community. A Church in which, week by week, services should be devoted, not to the iteration of abstract propositions in theology, but to the setting before men's minds of an ideal of true, just, and pure living ; a place in which those who are weary of the burden of daily cares should find a moment's rest in the contemplation of the higher life which is possible for all, though attained by so few ; a place in which the man of strife and of business should have time to think how small, after all, are the rewards he covets compared with peace and charity. Depend upon it, if such a Church existed, no one would seek to disestablish it.

The sole order of nobility which, in my judgment, becomes a philosopher, is the rank which he holds in the estimation of his fellow-workers, who are the only competent judges in such matters. Newton and Cuvier lowered themselves when the one accepted an idle knighthood, and the other became a baron of the empire. The great men who went to their graves as Michael Faraday and George Grote seem to me to have understood the

---

[1] This bears on his speech against Ultramontanism. (See p. 40).

dignity of knowledge better when they declined all such meretricious trappings. [1]

The usual note of high pressure recurs in the following letter, written to thank Darwin for his new work, *The Descent of Man, and Sexual Selection.*

JERMYN STREET, *Feb.* 20, 1871.

MY DEAR DARWIN—Best thanks for your new book, a copy of which I find awaiting me this morning. But I wish you would not bring your books out when I am so busy with all sorts of things. You know I can't show my face anywhere in society without having read them— and I consider it too bad.

No doubt, too, it is full of suggestions just like that I have hit upon by chance at p. 212 of vol. i., which connects the periodicity of vital phenomena with antecedent conditions.

Fancy lunacy, etc., coming out of the primary fact that one's nth ancestor lived between tide-marks ! I declare it's the grandest suggestion I have heard of for an age.

I have been working like a horse for the last fortnight, with the fag end of influenza hanging about me—and I am improving under the process, which shows what a good tonic work is.

I shall try if I can't pick out from " Sexual Selection " some practical hint for the improvement of gutter-babies, and bring in a resolution thereupon at the School Board. —Ever yours faithfully,   T. H. HUXLEY.

---

[1] On the other hand, he thought it right and proper for officials, in scientific as in other departments, to accept such honours, as giving them official power and status. In his own case, while refusing all simple titular honours, he accepted the Privy Councillorship, because, though incidentally carrying a title, it was an office ; and an office in virtue of which a man of science might, in theory at least, be called upon to act as responsible adviser to the Government, should special occasion arise.

This year also saw the inception of a scheme for a series of science primers, under the joint editorship of Professors Huxley, Roscoe, and Balfour Stewart. Huxley undertook the Introductory Primer, but it progressed slowly owing to pressure of other work, and was not actually finished till 1880.

26 ABBEY PLACE, *June* 29, 1871.

MY DEAR ROSCOE—If you could see the minutes of the Proceedings of the Aid to Science Commission, the Contagious Diseases Commission and the School Board (to say nothing of a lecture to Schoolmasters every morning), you would forgive me for not having written to you before.

But now that I have had a little time to look at it, I hasten to say that your chemical primer appears to me to be admirable—just what is wanted.

I enclose the sketch for my Primer *primus.* You will see the bearing of it, rough as it is. When it touches upon chemical matters, it would deal with them in a more rudimentary fashion than yours does, and only prepare the minds of the fledglings for you.

I send you a copy of the Report of the Education Committee, the resolutions based on which I am now slowly getting passed by our Board. The adoption of (*c*) among the essential subjects has, I hope, secured the future of Elementary Science in London. Cannot you get as much done in Manchester ?—Ever yours faithfully,

T. H. HUXLEY.

Sir Charles Lyell was now nearly 74 years old, and though he lived four years longer, age was beginning to tell even upon his vigorous powers. A chance meeting with him elicited the following letter :—

26 ABBEY PLACE, *July* 30, 1871.

MY DEAR DARWIN—I met Lyell in Waterloo Place to-day walking with Carrick Moore—and although what you said the other day had prepared me, I was greatly shocked at his appearance, and still more at his speech. There is no doubt it is affected in the way you describe, and the fact gives me very sad forebodings about him. The Fates send me a swift and speedy end whenever my time comes. I think there is nothing so lamentable as the spectacle of the wreck of a once clear and vigorous mind !

I am glad Frank enjoyed his visit to us. He is a great favourite here, and I hope he will understand that he is free of the house. It was the greatest fun to see Jess and Mady[1] on their dignity with him. No more kissing, I can tell you. Miss Mady was especially sublime.

Six out of our seven children have the whooping-cough. Need I say therefore that the wife is enjoying herself ?

With best regards to Mrs. Darwin and your daughter (and affectionate love to Polly) believe me—Ever yours faithfully, T. H. HUXLEY.

The purchase of the microscope, already referred to, was the subject of another letter to Dr. Dohrn, of which only the concluding paragraph about the School Board, is of general interest. Unfortunately the English microscope did not turn out a success, as compared to the work of the Jena opticians : this is the " optical Sadowa " of the second letter.

I fancy from what you wrote to my wife that there has been some report of my doings about the School Board in Germany. So I send you the number of the

---

[1] Aged 13 and 12 respectively.

Contemporary Review[1] for December that you may see
what line I have really taken.   Fanatics on both sides
abuse me, so I think I must be right.

When is this infernal war to come to an end ?   I hold
for Germany as always, but I wish she would make peace.
—With best wishes for the New Year—Ever yours,
                                    T. H. HUXLEY.

                              26 ABBEY PLACE, *July 7*, 1871.

MY DEAR DOHRN—I have received your packet, and
I will take care that your Report is duly presented to
the Association.   But the "Happy Family" in general,
and myself in particular, are very sorry you cannot come
to Scotland.   We had begun to count upon it, and the
children are immeasurably disgusted with the Insects
which will not lay their eggs at the right time.

You have become acclimatised to my bad behaviour in
the matter of correspondence, so I shall not apologise for
being in arrear.   I have been frightfully hard-worked
with two Royal Commissions and the School Board all
sitting at once, but I am none the worse, and things are
getting into shape—which is a satisfaction for one's
trouble.   I look forward hopefully towards getting back
to my ordinary work next year.

Your penultimate letter was very interesting to me,
but the glimpses into your new views which it affords
are very tantalising—and I want more.   What you say
about the development of the Amnion in your last letter
still more nearly brought "Donner und Blitz!" to my
lips—and I shall look out anxiously for your new facts.
Lankester tells me you have been giving lectures on your
views.   I wish I had been there to hear.

He is helping me as Demonstrator in a course of instruction in Biology which I am giving to Schoolmasters
—with the view of converting them into scientific

---

[1] Containing his article on "The School Boards," etc.

missionaries to convert the Christian Heathen of these islands to the true faith.

I am afraid that the English microscope turned out to be by no means worth the money and trouble you bestowed upon it. But the glory of such an optical Sadowa should count for something! I wish that you would get your Jena man to supply me with one of his best objectives if the price is not ruinous—I should like to compare it with my $\frac{1}{12}$ in. of Ross. [1]

All our children but Jessie have the whooping-cough —Pertussis—I don't know your German name for it. It is distressing enough for them, but, I think, still worse for their mother. However, there are no serious symptoms, and I hope the change of air will set them right.

They all join with me in best wishes and regrets that you are not coming. Won't you change your mind? We start on July 31st.—Ever yours faithfully,

T. H. HUXLEY.

The summer holiday of 1871 was spent at St. Andrews, a place rather laborious of approach at that time, with all the impedimenta of a large and young family, but chosen on account of its nearness to Edinburgh, where the British Association met that year. I well remember the night journey of some ten or eleven hours, the freshness of the early morning at Edinburgh, the hasty excursion with my father up the hill from the station as far as the old High Street. The return journey, however, was made easier by the kindness of Dr. Matthews Duncan, who put up the whole family for a night, so as to break the journey.

[1] In this connection it may be noted that he himself invented a combination microscope for laboratory use, still made by Crouch the optician. (See *Journ. Queckett Micr. Club*, vol. v. p. 144.)

We stayed at Castlemount, now belonging to Miss Paton, just opposite the ruined castle. Among other visitors to St. Andrews known to my father were Professors Tait and Crum Brown, who inveigled him into making trial of the "Royal and Ancient" game, which then, as now, was the staple resource of the famous little city. I have a vivid recollection of his being hopelessly bunkered three or four holes from home, and can testify that he bore the moral strain with more than usual calm as compared with the generality of golfers. Indeed, despite his naturally quick temper and his four years of naval service at a time when, perhaps, the traditions of a former generation had not wholly died out, he had a special aversion to the use of expletives; and the occasional appearance of a strong word in his letters must be put down to a simply literary use which he would have studiously avoided in conversation. A curious physical result followed the vigour with which he threw himself into the unwonted recreation. For the last twenty years his only physical exercise had been walking, and now his arms went black and blue under the muscular strain, as if they had been bruised.

But the holiday was by no means spent entirely in recreation. One week was devoted to the British Association; another to the examination of some interesting fossils at Elgin; while the last three weeks were occupied in writing two long articles, "Mr. Darwin's Critics," and the address entitled

"Administrative Nihilism" referred to above (p. 52), as well as a review of Dana's *Crinoids*. The former, which appeared in the *Contemporary Review* for November (*Coll. Ess.* ii. 120-187) was a review of (1) *Contributions to the Theory of Natural Selection*, by A. R. Wallace, (2) *The Genesis of Species*, by St. George Mivart, F.R.S., and (3) an article in the *Quarterly* for July 1871, on Darwin's *Descent of Man*.

"I am Darwin's bull-dog," he once said, and the *Quarterly Reviewer's* treatment of Darwin, "alike unjust and unbecoming," provoked him into immediate action. "I am about sending you," he writes to Haeckel on Nov. 2, "a little review of some of Darwin's critics. The dogs have been barking at his heels too much of late." Apart from this stricture, however, he notes the "happy change" which "has come over Mr. Darwin's critics. The mixture of ignorance and insolence which at first characterised a large proportion of the attacks with which he was assailed, is no longer the sad distinction of anti-Darwinian criticism." Notes too "that, in a dozen years, the *Origin of Species* has worked as complete a revolution in biological science as the *Principia* did in astronomy—and it has done so, because, in the words of Helmholtz, it contains 'an essentially new creative thought.'"

The essay is particularly interesting as giving evidence of his skill and knowledge in dealing with psychology, as against the *Quarterly Reviewer*, and even with such an unlikely subject as scholastic metaphysics, so that, by an odd turn of events, he appeared in the novel character of a defender of Catholic orthodoxy against an attempt from within that Church to prove that its teachings have in reality always been in harmony with the requirements of modern science.   For Mr. Mivart, while twitting the generality of men of science with their ignorance of the real doctrines of his church, gave a reference to the Jesuit theologian Suarez, the latest great representative of scholasticism, as following St. Augustine in asserting, not direct, but derivative creation, that is to say, evolution from primordial matter endued with certain powers.   Startled by this statement, Huxley investigated the works of the learned Jesuit, and found not only that Mr. Mivart's reference to the Metaphysical Disputations was not to the point, but that in the "Tractatus de opere sex Dierum," Suarez expressly and emphatically rejects this doctrine and reprehends Augustine for asserting it.

By great good luck (he writes to Darwin from St. Andrews) there is an excellent library here, with a good copy of Suarez, in a dozen big folios.   Among these I dived, to the great astonishment of the librarian, and looking into them as "the careful robin eyes the delver's toil" (vide Idylls), I carried off the two venerable clasped volumes which were most promising.

So I have come out in the new character of a defender of Catholic orthodoxy, and upset Mivart out of the mouth of his own prophet.

Darwin himself was more than pleased with the article, and wrote enthusiastically (see *Life and*

Letters, iii. 148-150). A few of his generous words may be quoted to show the rate at which he valued his friend's championship.

What a wonderful man you are to grapple with those old metaphysico-divinity books. . . . The pendulum is now swinging against our side, but I feel positive it will soon swing the other way; and no mortal man will do half as much as you in giving it a start in the right direction, as you did at the first commencement.

And again, after "mounting climax on climax," he continues :—" I must tell you what Hooker said to me a few years ago. 'When I read Huxley, I feel quite infantile in intellect.'"

This sketch of what constituted his holiday—and it was not very much busier than many another holiday—may possibly suggest what his busy time must have been like.

Till the end of the year the immense amount of work did not apparently tell upon him. He rejoiced in it. In December he remarked to his wife that with all his different irons in the fire, he had never felt his mind clearer or his vigour greater. Within a week he broke down quite suddenly, and could neither work nor think. He refers to this in the following letter :—

JERMYN STREET, *Dec.* 22, 1871.

MY DEAR JOHNNY—You are certainly improving. As a practitioner in the use of cold steel myself, I have read your letter in to-day's *Nature,* " mit Ehrfurcht und Bewunderung." And the best evidence of the greatness of your achievement is that it extracts this expression of admiration from a poor devil whose brains and body are in a colloid state, and who is off to Brighton for a day or two this afternoon.

God be with thee, my son, and strengthen the contents of thy gall-bladder !—Ever thine,   T. H. HUXLEY.

P.S.—Seriously, I am glad that at last a protest has been raised against the process of anonymous self-praise to which our friend is given.   I spoke to Smith the other day about that dose of it in the " *Quarterly* " article on Spirit-rapping.

CHAPTER III

1872

DYSPEPSIA, that most distressing of maladies, had laid firm hold upon him. He was compelled to take entire rest for a time. But his first holiday produced no lasting effect, and in the summer he was again very ill. Then the worry of a troublesome lawsuit in connection with the building of his new house intensified both bodily illness and mental depression. He had great fears of being saddled with heavy costs at the moment when he was least capable of meeting any new expense—hardly able even to afford another much-needed spell of rest. But in his case, as in others, at this critical moment the circle of fellow-workers in science to whom he was bound by ties of friendship, resolved that he should at least not lack the means of recovery. In their name Charles Darwin wrote him the following letter, of which it is difficult to say whether it does more honour to him who sent it or to him who received it :—

66

Down, Beckenham, Kent,
April 23, 1873.

My dear Huxley—I have been asked by some of
your friends (eighteen in number) to inform you that
they have placed through Robarts, Lubbock & Company,
the sum of £2100 to your account at your bankers. We
have done this to enable you to get such complete rest as
you may require for the re-establishment of your health ;
and in doing this we are convinced that we act for the
public interest, as well as in accordance with our most
earnest desires. Let me assure you that we are all your
warm personal friends, and that there is not a stranger
or mere acquaintance amongst us. If you could have
heard what was said, or could have read what was, as I
believe, our inmost thoughts, you would know that we
all feel towards you, as we should to an honoured and
much loved brother. I am sure that you will return
this feeling, and will therefore be glad to give us the
opportunity of aiding you in some degree, as this will
be a happiness to us to the last day of our lives. Let
me add that our plan occurred to several of your friends
at nearly the same time and quite independently of one
another.—My dear Huxley, your affectionate friend,

CHARLES DARWIN.

It was a poignant moment. "What have I done
to deserve this?" he exclaimed. The relief from
anxiety, so generously proffered, entirely overcame
him; and for the first time, he allowed himself to
confess that in the long struggle against ill-health, he
had been beaten; but, as he said, only enough to
teach him humility.

His first trip in search of health was in 1872,
when he obtained two months' leave of absence, and

prepared to go to the Mediterranean. His lectures to women on Physiology at South Kensington were taken over by Dr. Michael Foster, who had already acted as his substitute in the Fullerian course of 1868. But even on this cruise after health he was not altogether free from business. The stores of biscuit at Gibraltar and Malta were infested with a small grub and its cocoons. Complaints to the home authorities were met by the answer that the stores were prepared from the purest materials and sent out perfectly free from the pest. Discontent among the men was growing serious, when he was requested by the Admiralty to investigate the nature of the grub and the best means of preventing its ravages. In the end he found that the biscuits were packed within range of stocks of newly arrived, unpurified cocoa, from which the eggs were blown into the stores while being packed, and there hatched out. Thereafter the packing was done in another place and the complaints ceased.

Jan. 3, 1872.

MY DEAR DOHRN—It is true enough that I am somewhat "erkränkt," though beyond general weariness, incapacity and disgust with things in general, I do not precisely know what is the matter with me.

Unwillingly, I begin to suspect that I overworked myself last year. Doctors talk seriously to me, and declare that all sorts of wonderful things will happen if I do not take some more efficient rest than I have had for a long time. My wife adds her quota of persuasion and admonition, until I really begin to think I must do something, if only to have peace.

What if I were to come and look you up in Naples, somewhere in February, as soon as my lectures are over ?

The " one-plate system " might cure me of my incessant dyspeptic nausea. A detestable grub—larva of *Ephestia elatella*—has been devouring Her Majesty's stores of biscuits at Gibraltar. I have had to look into his origin, history, and best way of circumventing him—and maybe I shall visit Gibraltar and perhaps Malta. In that case, you will see me turn up some of these days at the Palazzo Torlonia.

Herbert Spencer has written a friendly attack on "Administrative Nihilism," which I will send you ; in the same number of the *Fortnightly* there is an absurd epicene splutter on the same subject by Mill's step-daughter, Miss Helen Taylor. I intended to publish the paper separately, with a note about Spencer's criticism, but I have had no energy nor faculty to do anything lately.

Tell Lankester, with best regards, that I believe the teaching of teachers in 1872 is arranged, and that I shall look for his help in due course.

The " Happy family " have had the measles since you saw them, but they are well again.

I write in Jermyn Street, so they cannot send messages ; otherwise there would be a chorus from them and the wife of good wishes and kind remembrances.— Ever yours,   T. H. HUXLEY.

He left Southampton on January 11, in the *Malta*. On the 16th, he notes in his diary, "I was up just in time to see the great portal of the Mediterranean well. It was a lovely morning, and nothing could be grander than Ape Hill on one side and the Rock on the other, looking like great lions or sphinxes on each side of a gateway."

The morning after his arrival he breakfasted with

Admiral Hornby, who sent him over to Tangier in the *Helicon*, giving the Bishop of Gibraltar a passage at the same time.   This led him to note down, "How the naval men love Baxter and all his works." A letter from Dr. Hooker to Sir John Hay ensured him a most hospitable welcome, though continual rain spoiled his excursions.   On the 21st he returned to Gibraltar, leaving three days later in the *Nyanza* for Alexandria, which was reached on February 1. At that "muddy hole" he landed in pouring rain, and it was not till he reached Cairo the following day that he at last got into his longed-for sunshine.

Seeing that three of his eight weeks had been spent in merely getting to sunshine, his wife and doctor conspired to apply for a third month of leave, which was immediately granted, so that he was able to accept the invitation of two friends to go with them up the Nile as far as Assouan in that most restful of conveyances, a dahabieh.

Cairo more than answered his expectations.   He stayed here till the 13th, making several excursions in company with Sir W. Gregory, notably to Boulak Museum, where he particularly notes the "man with ape" from Memphis; and, of course, the pyramids, of which he remarks that Cephren's is cased at the top with limestone, not granite.   His notebook and sketch-book show that he was equally interested in archæology, in the landscape and scenes of everyday life, and in the peculiar geographical and geological features of the county.   His first impression of the

Delta was its resemblance to Belgium and Lincoln-shire. He has sections and descriptions of the Mokatta hill, and the windmill mound, with a general panorama of the surrounding country and an explana-tion of it. He remarks at Memphis how the unburnt brick of which the mounds are made up had in many places become *remanie* into a stratified deposit—dis-tinguishable from Nile mud chiefly by the pottery fragments—and notes the bearing of this fact on the Cairo mounds. It is the same on his trip up the Nile; he jots down the geology whenever opportunity offered; remarks, as indication of the former height of the river, a high mud-bank beyond Edfou, and near Assouan a pot-hole in the granite fifty feet above the present level. Here is a detailed descrip-tion of the tomb of Aahmes; there a river-scene beside the pyramid of Meidum; or vivid sketches of vulture and jackal at a meal in the desert, the jackal in possession of the carcass, the vulture impatiently waiting his good pleasure for the last scraps; of the natives working at the endless shadoofs; of a group of listeners around a professional story-teller—un-finished, for he was observed sketching them.

Egypt left a profound impression upon him. His artistic delight in it apart, the antiquities and geology of the country were a vivid illustration to his trained eye of the history of man and the influence upon him of the surrounding country, the link between geography and history.

He left behind him for a while a most unexpected

memorial of his visit. A friend not long after going to the pyramids, was delighted to find himself thus adjured by a donkey-boy, who tried to cut out his rival with "Not him donkey, sah; him donkey bad, sah; my donkey good; my donkey 'Fessor-uxley donkey, sah." It appears that the Cairo donkey-boys have a way of naming their animals after celebrities whom they have borne on their backs.

While at Thebes, on his way down the river again, he received news of the death of the second son of Matthew Arnold, to whom he wrote the following letter :—

THEBES, *March* 10, 1872.

MY DEAR ARNOLD—I cannot tell you how shocked I was to see in the papers we received yesterday the announcement of the terrible blow which has fallen upon Mrs. Arnold and yourself.

Your poor boy looked such a fine manly fellow the last time I saw him, when we dined at your house, that I had to read the paragraph over and over again before I could bring myself to believe what I read. And it is such a grievous opening of a wound hardly yet healed that I hardly dare to think of the grief which must have bowed down Mrs. Arnold and yourself.

I hardly know whether I do well in writing to you. If such trouble befell me there are very few people in the world from whom I could bear even sympathy—but you would be one of them, and therefore I hope that you will forgive a condolence which will reach you so late as to disturb rather than soothe, for the sake of the hearty affection which dictates it.

My wife has told me of the very kind letter you wrote her. I was thoroughly broken down when I left England, and did not get much better until I fell into the utter and absolute laziness of dahabieh life. A month of that has completely set me up. I am as well as ever ; and though very grateful to Old Nile for all that he has done for me—not least for a whole universe of new thoughts and pictures of life—I begin to feel strongly

'the need of a world of men for me.'

But I am not going to overwork myself again. Pray make my kindest remembrances to Mrs. Arnold, and believe me, always yours very faithfully,

T. H. HUXLEY.

Leaving Assouan on March 3, and Cairo on the 18th, he returned by way of Messina to Naples, taking a day at Catania to look at Etna. At Naples he found his friend Dohrn was absent, and his place as host was filled by his father. Vesuvius was ascended, Pozzuoli and Pompeii visited, and two days spent in Rome.

HOTEL DE GRANDE BRETAGNE, NAPLES,
March 31, 1872.

MY DEAR TYNDALL—Your very welcome letter did not reach me until the 18th of March, when I returned to Cairo from my expedition to Assouan. Like Johnny Gilpin, I "little thought, when I set out, of running such a rig"; but while at Cairo I fell in with Ossory of the Athenæum, and a very pleasant fellow, Charles Ellis, who had taken a dahabieh, and were about to start up the Nile. They invited me to take possession of a vacant third cabin, and I accepted their hospitality, with the intention of going as far as Thebes and returning on my own hook. But when we got to Thebes I found there was no getting away again without much more exposure and fatigue than I felt justified in facing just then, and as my friends showed no disposition to be rid of me, I stuck to the boat, and only left them on the return voyage at Rodu, which is the terminus of the railway, about 150 miles from Cairo.

We had an unusually quick journey, as I was little more than a month away from Cairo, and as my companions made themselves very agreeable, it was very pleasant. I was not particularly well at first, but by degrees the utter rest of this "always afternoon" sort of life did its work, and I am as well and vigorous now as ever I was in my life.

I should have been home within a fortnight of the time I had originally fixed. This would have been ample time to have enabled me to fulfil all the engagements I had made before starting; and Donnelly had given me to understand that "My Lords" would not trouble their heads about my stretching my official leave. Nevertheless I was very glad to find the official extension (which was the effect of my wife's and your and Bence Jones's friendly conspiracy) awaiting me at Cairo. A rapid journey home *viâ* Brindisi might have rattled my brains back into the colloid state in which they were when I left England. Looking back through the past six months I begin to see that I have had a narrow escape from a bad breakdown, and I am full of good resolutions.

As the first-fruit of these you see that I have given up the School Board, and I mean to keep clear of all that semi-political work hereafter. I see that Sandon (whom I met at Alexandria) and Miller have followed my example, and that Lord Lawrence is likely to go. What a skedaddle!

It seems very hard to escape, however. Since my arrival here, on taking up the *Times* I saw a paragraph about the Lord Rectorship of St. Andrews. After enumerating a lot of candidates for that honour, the paragraph concluded, "But we understand that at present Professor Huxley has the best chance." It is really too bad if any one has been making use of my name without my permission. But I don't know what to do about it.

I had half a mind to write to Tulloch to tell him that I can't and won't take any such office, but I should look rather foolish if he replied that it was a mere newspaper report, and that nobody intended to put me up.

Egypt interested me profoundly, but I must reserve the tale of all I did and saw there for word of mouth. From Alexandria I went to Messina, and thence made an excursion along the lovely Sicilian coast to Catania and Etna. The old giant was half covered with snow, and this fact, which would have tempted you to go to the top, stopped me. But I went to the Val del Bove, whence all the great lava streams have flowed for the last two centuries, and feasted my eyes with its rugged grandeur. From Messina I came on here, and had the great good fortune to find Vesuvius in eruption. Before this fact the vision of good Bence Jones forbidding much exertion vanished into thin air, and on Thursday up I went in company with Ray Lankester and my friend Dohrn's father, Dohrn himself being unluckily away. We had a glorious day, and did not descend till late at night. The great crater was not very active, and contented itself with throwing out great clouds of steam and volleys of red-hot stones now and then. These were thrown towards the south-west side of the cone, so that it was practicable to walk all round the northern and eastern lip, and look down into the Hell Gate. I wished you were there to enjoy the sight as much as I did. No lava was issuing from the great crater, but on the north side of this, a little way below the top, an independent cone had established itself as the most charming little pocket-volcano imaginable. It could not have been more than 100 feet high, and at the top was a crater not more than six or seven feet across. Out of this, with a noise exactly resembling a blast furnace and a slowly-working high pressure steam engine combined, issued a violent torrent of steam and fragments of semi-fluid lava as big as one's fist, and sometimes bigger. These shot up sometimes as much as

100 feet, and then fell down on the sides of the little crater, which could be approached within fifty feet without any danger. As darkness set in, the spectacle was most strange. The fiery stream found a lurid reflection in the slowly-drifting steam cloud, which overhung it, while the red-hot stones which shot through the cloud shone strangely beside the quiet stars in a moonless sky.

Not from the top of this cinder cone, but from its side, a couple of hundred feet down, a stream of lava issued. At first it was not more than a couple of feet wide, but whether from receiving accessions or merely from the different form of slope, it got wider on its journey down to the Atrio del Cavallo, a thousand feet below. The slope immediately below the exit must have been near fifty, but the lava did not flow quicker than very thick treacle would do under like circumstances. And there were plenty of freshly cooled lava streams about, inclined at angles far greater than those which that learned Academician, Elie de Beaumont, declared to be possible. Naturally I was ashamed of these impertinent lava currents, and felt inclined to call them "Laves mousseuses."[1]

Courage, my friend, behold land! I know you love my handwriting. I am off to Rome to-day, and this day-week, if all goes well, I shall be under my own roof-tree again. In fact I hope to reach London on Saturday evening. It will be jolly to see your face again.—Ever yours faithfully,　　　　　　　T. H. HUXLEY.

My best remembrances to Hirst if you see him before I do.

My father reached home on April 6, sunburnt and bearded almost beyond recognition, but not really well, for as soon as he began work again in London,

---

[1] Elie de Beaumont "is said to have 'damned himself to everlasting fame' by inventing the nickname of 'la science moussante' for Evolutionism." See Life of Darwin, ii. 185.

his old enemy returned. Early hours, the avoidance
of society and societies, an hour's riding before starting at nine for South Kensington, were all useless ;
the whole year was poisoned until a special diet
prescribed by Dr. (afterwards Sir) Andrew Clark,
followed by another trip abroad, effected a cure. I
remember his saying once that he learned by sad
experience that such a holiday as that in Egypt was
no good for him. What he really required was
mountain air and plenty of exercise. The following
letters fill up the outline of this period :—

26 ABBEY PLACE, May 20, 1872.

MY DEAR DOHRN—I suppose that you are now back
in Naples, perambulating the Chiaja, and looking ruefully
on the accumulation of ashes on the foundations of the
aquarium ! The papers, at any rate, tell us that the
ashes of Vesuvius have fallen abundantly at Naples.
Moreover, that abominable municipality is sure to have
made the eruption an excuse for all sorts of delays. May
the gods give you an extra share of temper and patience !

What an unlucky dog our poor Ray is, to go and get
fever when of all times in the world's history he should
not have had it. However, I hear he is better and on
his way home. I hope he will be well enough when he
returns not only to get his Fellowship, but to help me in
my schoolmaster work in June and July.

I was greatly disgusted to miss you in Naples, but it
was something to find your father instead. What a
vigorous, genial youngster of threescore and ten he is. I
declare I felt quite aged beside him. We had a glorious
day on Vesuvius, and behaved very badly by leaving him
at the inn for I do not know how many hours, while we
wandered about the cone. But he had a very charming
young lady for companion, and possibly had the best of

it. I am very sorry that at the last I went off in a hurry without saying "Good-bye" to him, but I desired Lankester to explain, and I am sure he will have sympathised with my anxiety to see Rome.

I returned, thinking myself very well, but a bad fit of dyspepsia seized me, and I found myself obliged to be very idle and very careful of myself—neither of which things are to my taste. But I am right again now, and hope to have no more backslidings. However, I am afraid I may not be able to attend the Brighton meeting. In which case you will have to pay us a visit, wherever we may be—where, we have not yet made up our minds, but it will not be so far as St. Andrews.

Now for a piece of business. The new Governor of Ceylon is a friend of mine, and is proposing to set up a Natural History Museum in Ceylon. He wants a curator —some vigorous fellow with plenty of knowledge and power of organisation who will make use of his great opportunities. He tells me he thinks he can start him with £350 a year (and a house) with possible increase to £400. I do not know any one here who would answer the purpose. Can you recommend me any one? If you can, let me know at once, and don't take so long in writing to me as I have been in writing to you.

I await the "Prophecies of the Holy Antonius"[1] anxiously. Like the Jews of old, I come of an unbelieving generation, and need a sign. The bread and the oil, also the chamber in the wall shall not fail the prophet when he comes in August : nor Donner and Blitzes either.

I leave the rest of the space for the wife.—Ever yours,
T. H. H.

The following is in reply to a jest of Dr. Dohrn's —who was still a bachelor—upon a friend's unusual sort of offering to a young lady.

---

[1] His work on the development of the Arthropoda or Spider family.

I suspected the love affair you speak of, and thought the young damsel very attractive. I suppose it will come to nothing, even if he be disposed to add his hand to the iron and quinine, in the next present he offers. . . . And, oh my Diogenes, happy in a tub of arthropodous Entwickelungsgeschichte,[1] despise not beefsteaks, nor wives either. They also are good.

JERMYN STREET, *June 5*, 1872.

MY DEAR DOHRN—I have written to the Governor of Ceylon, and enclosed the first half of your letter to me to him as he understands High Dutch. I have told him that the best thing he can do is to write to you at Naples and tell you he will be very happy to see you as soon as you can come. And that if you do come you will give him the best possible advice about his museum, and let him have no rest until he has given you a site for a zoological station.

I have no doubt you will get a letter from him in three weeks or so. His name is Gregory, and you will find him a good-humoured acute man of the world, with a very great general interest in scientific and artistic matters. Indeed in art I believe he is a considerable connoisseur.

I am very grieved to hear of your father's serious illness. At his age cerebral attacks are serious, and when we spent so many pleasant hours together at Naples, he seemed to have an endless store of vigour—very much like his son Anton.

What put it into your head that I had any doubt of your power of work? I am ready to believe that you are Hydra in the matter of heads and Briareus in the matter of hands.

. . If you go to Ceylon I shall expect you to come back by way of England. It's the shortest route any-where from India, though it may not look so on the map.

---

[1] *History of Development.*

How am I? Oh, getting along and just keeping the devil of dyspepsia at arm's length. The wife and other members of the H. F. are well, and would send you greetings if they knew I was writing to you.—Ever yours faithfully,   T. H. HUXLEY.

A little later Von Willemoes Suhm ("why the deuce does he have such a long name, instead of a handy monosyllable or dissyllable like Dohrn or Huxley?") was recommended for the post. He afterwards was one of the scientific staff of the *Challenger*, and died during the voyage.

MORTHOE, NEAR BARNSTAPLE, NORTH DEVON,
Aug. 5, 1872.

MY DEAR DOHRN—I trust you have not been very wroth with me for my long delay in answering your last letter. For the last six weeks I have been very busy lecturing daily to a batch of schoolmasters, and looking after their practical instruction in the laboratory which the Government has, at last, given me. In the "intervals of business" I have been taking my share in a battle which has been raging between my friend Hooker of Kew and his official chief. . . . And moreover I have just had strength enough to get my daily work done and no more, and everything that could be put off has gone to the wall. Three days ago, the "Happy Family," bag and baggage, came to this remote corner, where I propose to take a couple of months' entire rest—and put myself in order for next winter's campaign. It is a little village five miles from the nearest town (which is Ilfracombe), and our house is at the head of a ravine running down to the sea. Our backs are turned to England and our faces to America with no land that I know of between. The country about is beautiful, and if you will come we will put you up at the little inn, and show you something better than even Swanage. There are slight difficulties about the commissariat, but that is the Hausfrau's business, and not mine. At the worst, bread, eggs, milk, and rabbits are certain, and the post from London takes two days!

MORTHOE, ILFRACOMBE, N. DEVON,
Aug. 23, 1872.

MY DEAR WHIRLWIND—I promise you all my books, past, present, and to come for the Aquarium. The best part about them is that they will not take up much room. Ask for Owen's by all means; "Fas est etiam ab hoste doceri." I am very glad you have got the British Association publications, as it will be a good precedent for the Royal Society.

Have you talked to Hooker about marine botany? He may be able to help you as soon as X. the accursed (may jackasses sit upon his grandmother's grave, as we say in the East) leaves him alone.

It is hateful that you should be in England without seeing us, and for the first time I lament coming here. The children howled in chorus when they heard that you could not come. At this moment the whole tribe and their mother have gone to the sea, and I must answer your letter before the post goes out, which it does here about half an hour after it comes in.—Ever yours very faithfully,   T. H. HUXLEY.

In 1872 Huxley was at length enabled to establish in his regular classes a system of science teaching based upon laboratory work by the students, which he had long felt to be the only true method. It involved the verification of every fact by each student, and was a training in scientific method even more than in scientific fact. Had circumstances only permitted, the new epoch in biological teaching

might have been antedated by many years.   But, as
he says in the preface to the *Practical Biology*, 1875—

Practical work was forbidden by the limitations of
space in the building in Jermyn Street, which possessed
no room applicable to the purpose of a laboratory, and I
was obliged to content myself, for many years, with what
seemed the next best thing, namely, as full an exposition
as I could give of the characters of certain plants and
animals, selected as types of vegetable and animal organisation, by way of introduction to systematic zoology and
paleontology.

There was no laboratory work, but he would show
an experiment or a dissection during the lecture or
perhaps for a few minutes after, when the audience
crowded round the lecture table.

The opportunity came in 1871.   As he afterwards
impressed upon the great city companies in regard to
technical education, the teaching of science throughout the country turned upon the supply of trained
teachers.   The part to be played by elementary
science under the Education Act of 1870, added
urgency to the question of proper teaching.   With
this in view, he organised a course of instruction for
those who had been preparing pupils for the examinations of the Science and Art Department, " scientific
missionaries," as he described them to Dr. Dohrn.

In the promotion of the practical teaching of biology
(writes the late Jeffery Parker, *Nat. Sci.* viii. 49), Huxley's
services can hardly be overestimated.   Botanists had
always been in the habit of distributing flowers to their
students, which they could dissect or not as they chose;

animal histology was taught in many colleges under the
name of practical physiology; and at Oxford an excellent
system of zoological work had been established by the late
Professor Rolleston.[1]   But the biological laboratory, as it
is now understood, may be said to date from about 1870,
when Huxley, with the co-operation of Professors Foster,
Rutherford, Lankester, Martin, and others,[2] held short
summer classes for science teachers at South Kensington,
the daily work consisting of an hour's lecture followed by
four hours' laboratory work, in which the students verified
for themselves facts which they had hitherto heard about
and taught to their unfortunate pupils from books alone.
The naïve astonishment and delight of the more intelligent
among them was sometimes almost pathetic.   One clergy-
man, who had for years conducted classes in physiology
under the Science and Art Department, was shown a drop
of his own blood under the microscope.   " Dear me ! "
he exclaimed, " it's just like the picture in Huxley's
Physiology."

---

[1] " Rolleston (Professor Lankester writes to me) was the first to
systematically conduct the study of Zoology and Comparative
Anatomy in this country by making use of a carefully selected
series of animals.   His ' types ' were the Rat, the Common Pigeon,
the Frog, the Perch, the Crayfish, Blackbeetle, Anodon, Snail,
Earthworm, Leech, Tapeworm.   He had a series of dissections of
these mounted, also loose dissections and elaborate MS. descriptions.
The student went through this series, dissecting fresh specimens for
himself.   After some ten years' experience Rolleston printed his
MS. directions and notes as a book, called *Forms of Animal Life.*

"This all preceded the practical class at South Kensington in
1871.   I have no doubt that Rolleston was influenced in his plan
by your father's advice.   But Rolleston had the earlier opportunity
of putting the method into practice.

" Your father's series of types were chosen so as to include
plants, and he gave more attention to microscopic forms and to
microscopic structure than did Rolleston."

It was distinctive of the lectures that they were on biology, on
plants as well as animals, to illustrate all the fundamental features
of living things.

[2] T. J. Parker, G. B. Howes, and the present Sir W. Thiselton
Dyer, K.C.M.G., C.I.E.

Later, in 1872, when the biological department of
the Royal School of Mines was transferred to South
Kensington, this method was adopted as part of the
regular curriculum of the school, and from that time
the teaching "of zoology by lectures alone became an
anachronism."

The first of these courses to schoolmasters took
place, as has been said, in 1871. Some large rooms
on the ground floor of the South Kensington Museum
were used for the purpose. There was no proper
laboratory, but professor and demonstrators rigged
up everything as wanted. Huxley was in the full
tide of that more than natural energy which preceded
his breakdown in health, and gave what Professor
Ray Lankester describes as " a wonderful course of
lectures," one every day from ten to eleven for six
weeks, in June and half July. The three demonstrators (those named first on the list above) each took
a third of the class, about thirty-five apiece. "Great
enthusiasm prevailed. We went over a number of
plants and of animals—including microscopic work
and some physiological experiment. The 'types'
were more numerous than in later courses."

In 1872 the new laboratory—the present one—
was ready. "I have a laboratory," writes Huxley to
Dohrn, "which it shall do your eyes good to behold
when you come back from Ceylon, the short way " (i.e.
viâ England). Here a similar course, under the same
demonstrators, assisted by H. N. Martin, was given
in the summer, Huxley, though very shaky in

health, making a point of carrying them out himself.

26 ABBEY PLACE, June 4, 1872.

MY DEAR TYNDALL—I must be at work on examination
papers all day to-day, but to-morrow I am good to lunch
with you (and abscond from the Royal Commission,
which will get on very well without me) or to go with
you and call on your friends, whichever may be most
convenient.

Many thanks for all your kind and good advice about
the lectures, but I really think they will not be too
much for me, and it is of the utmost importance I should
carry them on.

They are the commencement of a new system of
teaching which, if I mistake not, will grow into a big
thing and bear great fruit, and just at this present
moment (nobody is necessary very long) I am the necessary
man to carry it on. I could not get a suppléant if I
would, and you are no more the man than I am to let a
pet scheme fall through for the fear of a little risk of self.
And really and truly I find that by taking care I pull
along very well. Moreover, it isn't my brains that get
wrong, but only my confounded stomach.

I have read your memorial [1] which is very strong and
striking, but a difficulty occurs to me about a good deal
of it, and that is that it won't do to quote Hooker's
official letters before they have been called for in Parliament, or otherwise made public. We should find ourselves in the wrong officially, I am afraid, by doing so.
However we can discuss this when we meet. I will be
at the Athenæum at 4 o'clock.—Ever yours faithfully,

T. H. HUXLEY.

As for the teaching by "types," which was the
most salient feature of his method, and therefore the

---

[1] In the affair of Dr. Hooker already referred to, p. 80.

most easily applied and misapplied, Professor Parker
continues :—

> Huxley's method of teaching was based upon the
> personal examination by the student of certain "types"
> of animals and plants selected with a view of illustrating
> the various groups. But, in his lectures, these types
> were not treated as the isolated things they necessarily
> appear in a laboratory manual or an examination syllabus ;
> each, on the contrary, took its proper place as an example
> of a particular grade of structure, and no student of
> ordinary intelligence could fail to see that the types
> were valuable, not for themselves, but simply as marking,
> so to speak, the chapters of a connected narrative. More-
> over, in addition to the types, a good deal of work of a
> more general character was done. Thus, while we owe
> to Huxley more than to any one else the modern system
> of teaching biology, he is by no means responsible for the
> somewhat arid and mechanical aspect it has assumed in
> certain quarters.

The application of the same system to botanical
teaching was inaugurated in 1873, when, being compelled to go abroad for his health, he arranged that
Mr. (now Sir W.) Thiselton Dyer should take his
place and lecture on Botany.

The *Elementary Instruction in Biology*, published in
1875, was a text-book based upon this system. This
book, in writing which Huxley was assisted by his
demonstrator, H. N. Martin, was reprinted thirteen
times before 1888, when it was "Revised and
Extended by Howes and Scott," his later assistants.
The revised edition is marked by one radical change,
due to the insistence of his demonstrator, the late

Prof. Jeffery Parker. In the first edition, the lower
forms of life were first dealt with; from simple cells
—amœba, yeast-plant, blood-corpuscle—the student
was taken through an ascending series of plants
and of animals, ending with the. frog or rabbit.
But "the experience of the Lecture-room and the
Laboratory taught me," writes Huxley in the new
preface, "that philosophical as it might be in theory,
it had defects in practice." The process might be
regarded as not following the scientific rule of proceeding from the known to the unknown; while the
small and simple organisms required a skill in hand-
ling high-power microscopes which was difficult for
beginners to acquire. Hence the course was reversed,
and began with the more familiar type of the rabbit
or frog. This was Rolleston's practice; but it may
be noted that Professor Ray Lankester has always
maintained and further developed "the original
Huxleian plan of beginning with the same microscopic forms" as being a most important philosophic
improvement on Rolleston's plan, and giving, he
considers, "the truer 'twist,' as it were, to a student's
mind."

When the book was sent to Darwin, he wrote
back (November 12, 1875) :—

MY DEAR HUXLEY—Many thanks for your biology,
which I have read. It was a real stroke of genius to
think of such a plan. Lord, how I wish that I had
gone through such a course.—Ever yours,

C. DARWIN.

A large portion of his time and energy was occupied in the organisation of this course of teaching for teachers, and its elaboration before being launched on a larger scale in October, when the Biological Department of the Jermyn Street school was transferred to the new buildings at South Kensington, fitted with laboratories which were to excite his friend Dr. Dohrn's envy. But he was also at work upon his share of the *Science Primers*, so far as his still uncertain health allowed. This and the affairs of the British Association are the subject of several letters to Sir Henry Roscoe and Dr. Tyndall.

26 ABBEY PLACE, *April* 8, 1872.

MY DEAR ROSCOE—Many thanks for your kind letter of welcome. My long rest has completely restored me. As my doctor told me, I was sound, wind and limb, and had merely worn myself out. I am not going to do that again, and you see that I have got rid of the School Board. It was an awful incubus !

Oddly enough I met the Ashtons in the Vatican, and heard about your perplexities touching Oxford. I should have advised you to do as you have done. I think that you have a great piece of work to do at Owens College, and that you will do it. If you had gone to Oxford you would have sacrificed all the momentum you have gained in Manchester; and would have had to begin *de novo*, among conditions which, I imagine, it is very hard for a non-University man to appreciate and adjust himself to.

I like the look of the " Primers" (of which Macmillan has sent me copies to-day) very much, and shall buckle to at mine as soon as possible. I am very glad you did not wait for me. I remained in a very shaky condition up to the middle of March, and could do nothing.—Ever yours very faithfully,   T. H. HUXLEY.

The wife unites with me in kind regards to Mrs. Roscoe and yourself.

MORTHOE, ILFRACOMBE, N. DEVON,
Sept. 9, 1872.

MY DEAR TYNDALL—I was very glad to have news of you, and to hear that you are vigorous.

My outing hitherto has not been very successful, so far as the inward man is concerned at least, for the weather has been good enough. But I have been worried to death with dyspepsia and the hypochondriacal bedevilments that follow in its train, until I am seriously thinking of returning to town to see if the fine air of St. John's Wood (as the man says in *Punch*) won't enable me to recover from the effects of the country.

I wish I were going with you to Yankee Land, not to do any lecturing, God forbid! but to be a quiet spectator in a corner of the enthusiastic audiences. I am as lazy as a dog, and the rôle of looker-on would just suit me. However, I have a good piece of work to do in organising my new work at South Kensington.

I have just asked my children what message they have to send to you, and they send their love; very sorry they won't see you before you go, and hope you won't come back speaking through your nose!

I shall be in town this week or next, and therefore *shall* see you.—Ever yours faithfully,

T. H. HUXLEY.

26 ABBEY PLACE, *Sept.* 17, 1872.

MY DEAR ROSCOE—Your letter has followed me from Morthoe here. We had good enough weather in Devon —but my stay there was marred by the continuous

dyspepsia and concurrent hypochondriacal incapacity
At last, I could not stand it any longer, and came home
for "change of air," leaving the wife and chicks to follow
next week. By dint of living on cocoa and Revalenta,
and giving up drink, tobacco, and all other things that
make existence pleasant, I am getting better.

What was your motive in getting kicked by a horse?
I stopped away from the Association without that; and
am not sorry to have been out of the way of the X.
business. What is to become of the association if——
is to monopolise it? And then there was that scoundrel,
Louis Napoleon—to whom no honest man ought to
speak—gracing the scene. I am right glad I was out
of it.

I am at my wits' end to suggest a lecturer for you.
I wish I could offer myself, but I have refused everything
of that sort on the score of health; and moreover, I am
afraid of my wife!

What do you say to Ramsay? He lectures very well.
I have done nothing whatever to the Primer. Stewart
sent me Geikie's letter this morning, and I have asked
Macmillan to send Geikie the proofs of my Primer so far
as they go. We must not overlap more than can be
helped.

I have not seen Hooker yet since my return. While
all this row has been going on, I could not ask him to do
anything for us. And until X. is dead and d—d (officially
at any rate), I am afraid there will be little peace for
him.—Ever yours very faithfully,   T. H. HUXLEY.

Please remember me very kindly to Mrs. Roscoe.

In a letter of September 25 is a reference to the
way in which his increasing family had outgrown his
house in Abbey Place. Early in the preceding year,
he had come to the decision to buy a small house in

the same neighbourhood, and add to it so as to give
elbow-room to each and all of the family. This was
against the advice of his friend and legal adviser, to
whom he wrote announcing his decision, as follows.
The letter was adorned with a sketch of an absurd
cottage, "Ye House!" perched like a windmill on a
kind of pedestal, and with members of the family
painfully ascending a ladder to the upper story,
above the ominous legend, "Staircase forgotten."

                                        March 20, 1871.

MY DEAR BURTON—There is something delightfully
refreshing in rushing into a piece of practical work in
the teeth of one's legal adviser.

If the lease of a piece of ground whereon I am going
to build mine house come to you, will you see if it's all
right.—Yours wilfully,   T. H. HUXLEY.

This house, No. 4 Marlborough Place, stands on
the north side of that quiet street, close to its junction
with Abbey Road. It is next door to the Presbyterian
Church, on the other side of which again is a Jewish
synagogue. The irregular front of the house, with
the original cottage, white-painted and deep-eaved,
joined by a big porch to the new uncompromising
square face of yellow brick, distinguished only by its
extremely large windows, was screened from the road
by a high oak paling, and a well-grown row of young
lime-trees. Taken as a whole, it was not without
character, and certainly was unlike most London
houses. It was built for comfort, not beauty ; designed, within stringent limits as to cost, to give each

member of the family room to get away by himself
or herself if so disposed.   Moreover, the gain in space
made it more possible to see something of friends or
put up a guest, than in the small and crowded house
in Abbey Place.

A small garden lay in front of the house; a considerably larger garden behind, wherein the chief
ornament was then a large apple-tree, that never
failed to spread a cloud of blossom for my father's
birthday, the 4th of May.

Over the way, too, for many years we were faced
by a long garden full of blossoming pear-trees in
which thrushes and blackbirds sang and nested,
belonging to a desolate house in the Abbey Road,
which was tenanted by a solitary old man, supposed
to be a male prototype of Miss Havisham in *Great
Expectations.*

The move was accompanied by a unique and unpleasant experience.   A knavish fellow, living in a
cottage close to the foot of the garden, sought to
blackmail the newcomer, under threat of legal proceedings, alleging that a catchment well for surface
drainage had made his basement damp.   Unfortunately for his case, it could be shown that the pipes
had not yet been connected with the well, and when
he carried out his threat, he gained nothing from his
suit in Chancery and his subsequent appeal, except
some stinging remarks from Vice-Chancellor Malins.

I am afraid the brute is impecunious (wrote my father
after the first suit failed), and that I shall get nothing

out of him. So I shall have had three months' worry, and be fined £100 or so for being wholly and absolutely in the right.

Happily the man turned out to have enough means to pay the bulk of the costs ; but that was no compensation for the mental worry and consequent ill-health entailed from November to June.

The only amusing point in the whole affair was when the plaintiff's solicitors had the face to file an affidavit before the Vice-Chancellor himself in answer to his strictures upon the case, "about as regular a proceeding," reports Mr. Burton, "as for a middy to reply upon the Post Captain on his own quarter-deck."

The move was made in the third week of December (1872) amid endless rain and mud and with workmen still in the house. It was attended by one inconvenience. He writes to Darwin on December 20, 1872 :—

I am utterly disgusted at having only just received your note of Tuesday. But the fact is, there is a certain inconvenience about having *four* addresses as has been my case for the most part of this week, in consequence of our moving—and as I have not been to Jermyn Street before to-day, I have missed your note. I should run round to Queen Anne St. now on the chance of catching you, but I am bound here by an appointment.

One incident of the move, however, was more agreeable. Mr. Herbert Spencer took the opportunity of sending a New Year's gift for the new house, in the shape of a handsome clock, wishing, as he said,

"to express in some way more emphatic than by
words, my sense of the many kindnesses I have
received at your hands during the twenty years of
our friendship. Remembrance of the things you
have done in furtherance of my aims, and of the
invaluable critical aid you have given me, with so
much patience and at so much cost of time, has often
made me feel how much I owe you."

After a generous reference to occasions when the
warmth of debate might have betrayed him into
more vigorous expressions than he intended, he
concludes :—

But inadequately as I may ordinarily show it, you
will (knowing that I am tolerably candid) believe me
when I say that there is no one whose judgment on all
subjects I so much respect, or whose friendship I so highly
value.

It may be remembered that the 1872 address on
"Administrative Nihilism" led to a reply from the
pen of Mr. Spencer, as the champion of Individualism.
When my father sent him the volume in which this
address was printed, he wrote back a letter (Sept. 29,
1873) which is characterised by the same feeling. It
expresses his thanks for the book, "and many more
for the kind expression of feeling in the preface. If
you had intended to set an example to the Philistines
of the way in which controversial differences may be
maintained without any decrease of sympathy, you
could not have done it more perfectly."

In connection with the building of the house,

Tyndall had advanced a sum of money to his friend, and with his usual generosity, not only received interest with the greatest reluctance, but would have liked to make a gift of the principal. He writes, "If I remain a bachelor I will circumvent you—if not—not. It cleaves to me like dirt—and that is why you wish to get rid of it." To this he received answer :—

Feb. 26, 1873.

I am not to be deterred by any amount of bribery and corruption, from bringing you under the yoke of a "rare and radiant,"—whenever I discover one competent to undertake the ticklish business of governing you. I hope she will be "radiant,"—uncommonly "rare" she certainly will be !

Two years later this loan was paid off, with the following letter :—

4 MARLBOROUGH PLACE,
Jan. 11, 1875.

MY DEAR OLD SHYLOCK—My argosies have come in, and here is all that was written in the bond ! If you want the pound of flesh too, you know it is at your service, and my Portia won't raise that pettifogging objection to shedding a little blood into the bargain, which that other one did.—Ever yours faithfully,
T. H. HUXLEY.

On October 24 Miss Jex Blake wrote to him to ask his help for herself and the other women medical students at Edinburgh. For two years they had only been able to get anatomical teaching in a mixed class ; but wishing to have a separate class, at least for the

present, they had tried to arrange for one that session.
The late demonstrator at the Surgeons' Hall, who
had given them most of their teaching before, had
undertaken to teach this separate class, but was
refused recognition by the University Court, on the
ground that they had no evidence of his qualifications,
while refusing to let him prove his qualification by
examination. This the women students understood
to be an indirect means of suppressing their aspirations ; they therefore begged Huxley to examine
their instructor with a view to giving him a certificate
which should carry weight with the University
Court.

He replied :—

Oct. 28, 1872.

DEAR MADAM—While I fully sympathise with the
efforts made by yourself and others, to obtain for women
the education requisite to qualify them for medical
practice, and while I think that women who have the
inclination and the capacity to follow the profession of
medicine are most unjustly dealt with if any obstacles
beyond those which are natural and inevitable are placed
in their way, I must nevertheless add, that I as completely
sympathise with those Professors of Anatomy, Physiology,
and Obstetrics, who object to teach such subjects to mixed
classes of young men and women brought together without
any further evidence of moral and mental fitness for such
association than the payment of their fees.

In fact, with rare exceptions, I have refused to admit
women to my own Lectures on Comparative Anatomy for
many years past. But I should not hesitate to teach
anything I know to a class composed of women ; and I
find it hard to believe that any one should really wish to

prevent women from obtaining efficient separate instruction, and from being admitted to Examination for degrees upon the same terms as men.

You will therefore understand that I should be most glad to help you if I could—and it is with great regret that I feel myself compelled to refuse your request to examine Mr. H——.

In the first place, I am in the midst of my own teaching, and with health not yet completely re-established I am obliged to keep clear of all unnecessary work. Secondly, such an examination must be practical, and I have neither dissecting-room available nor the anatomical license required for human dissection ; and thirdly, it is not likely that the University authorities would attach much weight to my report on one or two days' work—if the fact that Mr. H—— has already filled the office of anatomical Demonstrator (as I understand from you) does not satisfy them as to his competency.—I am, dear Madam, yours very faithfully,   T. H. HUXLEY.

Miss S. JEX BLAKE.

The last event of the year was that he was elected by the students Lord Rector of Aberdeen University —a position, the duties of which consist partly in attending certain meetings of the University Court, but more especially in delivering an address. This, however, was not required for another twelvemonth, and the address on " Universities, Actual and Ideal," was delivered in fulfilment of this duty in February 1874.

CHAPTER IV

1873

THE year opens with a letter to Tyndall, then on a lecturing tour in America :—

4 MARLBOROUGH PLACE, ABBEY ROAD, N.W.,
January 1, 1872 [1873].

MY DEAR TYNDALL—I cannot let this day go by without wishing you a happy New Year, and lamenting your absence from our customary dinner.   But Hirst and Spencer and Michael Foster are coming, and they shall drink your health in champagne while I do the like in cold water, making up by the strength of my good wishes for the weakness of the beverage.

You see I write from the new house.   Getting into it was an awful job, made worse than needful by the infamous weather we have had for weeks and months, and by the stupid delays of the workmen, whom we had fairly to shove out at last as we came in.   We are settling. down by degrees, and shall be very comfortable by and by, though I do not suppose that we shall be able to use the drawing-room for two or three months to come.   I am very glad to have made the change, but there is a drawback to everything in " this here wale," as Mrs. Gamp says, and my present thorn in the flesh is a neighbour, who says I have injured him by certain operations in my garden, and is trying to get something out of me by Chancery proceedings. Fancy finding myself a defendant in Chancery !

It is particularly hard on me, as I have been especially careful to have nothing done without Burton's sanction and assurance that I was quite safe in law ; and I would have given up anything [rather] than have got into bother of this kind. But "sich is life."

You seem to have been making a Royal Progress in Yankee-land. We have been uncommonly tickled with some of the reports of your lectures which reached us, especially with that which spoke of your having "a strong English accent."

The loss of your assistant seems to have been the only deduction to be made from your success. I am afraid you must have felt it much in all ways.

"My Lord" received your telegram only after the business of "securing Hirst" was done. That is one of the bright spots in a bad year for me. Goschen consulted Spottiswoode and me independently about the headship of the new Naval College, and was naturally considerably surprised by the fact that we coincided in recommending Hirst. . . . The upshot was that Goschen asked me to communicate with Hirst and see if he would be disposed to accept the offer. So I did, and found to my great satisfaction that Hirst took to the notion very kindly. I am sure he is the very best man for the post to be met with in the three kingdoms, having that rare combination of qualities by which he gets on with all manner of men, and singularly attracts young fellows. He will not only do his duty, but be beloved for doing it, which is what few people can compass.

I have little news to give you. The tail of the X.-Hooker storm is drifting over the scientific sky in the shape of fresh attacks by Owen on Hooker. Hooker answered the last angelically, and I hope they are at an end.

The wife has just come in and sends her love (but is careful to add "second-best"). The chicks grow visibly and audibly, and Jess looks quite a woman. All are well except myself, and I am getting better from a fresh breakdown of dyspepsia. I find that if I am to exist at all it must be on strictly ascetic principles, so there is hope of my dying in the odour of sanctity yet. If you recollect, Lancelot did not know that he should "die a holy man" till rather late in life. I have forgotten to tell you about the Rectorship of Aberdeen. I refused to stand at first, on the score of health, and only consented on condition that I should not be called upon to do any public work until after the long vacation. It was a very hard fight, and although I had an absolute majority of over fifty, the mode of election is such that one vote, in one of the four nations, would have turned the scale by giving my opponent the majority in that nation. We should then have been ties, and as the chancellor, who has under such circumstances a casting vote, would have (I believe) given it against me, I should have been beaten.

As it is, the fact of any one, who stinketh in the nostrils of orthodoxy, beating a Scotch peer at his own gates in the most orthodox of Scotch cities, is a curious sign of the times. The reason why they made such a tremendous fight for me is, I believe, that I may carry on the reforms commenced by Grant Duff, my predecessor. Unlike other Lord Rectors, he of Aberdeen is a power and can practically govern the action of the University during his tenure of office.

I saw Pollock yesterday, and he says that they want you back again. Curiously the same desire is epidemically prevalent among your friends, not least here.—Ever yours,   T. H. HUXLEY.

In spite of his anxieties, his health was slowly improving under careful regimen. He published no scientific memoirs this year, but in addition to his regular lectures, he was working to finish his *Manual of Invertebrate Anatomy* and his *Introductory Primer*, and to write his Aberdeen address; he was also at work upon the *Pedigree of the Horse* and on *Bodily Motion and Consciousness*. He delivered a course to teachers on Psychology and Physiology, and was much occupied by the Royal Commission on Science. As a governor of Owens College he had various meetings to attend, though his duties did not extend, as some of his friends seem to have thought, to the appointment of a Professor of Physiology there.

My life (he writes to Sir Henry Roscoe) is becoming a burden to me because of ——. Why I do not know, but for some reason people have taken it into their heads that I have something to do with appointments in Owens College, and no fewer than three men of whose opinion 1 think highly have spoken or written to me urging ——'s merits very strongly.

This summer he again took a long holiday, thanks to the generosity of his friends (see p. 67), and with better results. He went with his old friend Hooker to the Auvergne, walking, geologising, sketching, and gradually discarding doctor's orders. Sir Joseph Hooker has very kindly written me a letter from which I give an account of this trip :—

It was during the many excursions we took together, either by ourselves or with one of my boys, that I knew him best at his best; and especially during one of several weeks' duration in the summer of 1873, which we spent in central France and Germany. He had been seriously ill, and was suffering from severe mental depression. For this he was ordered abroad by his physician, Sir A. Clark, to which step he offered a stubborn resistance. With Mrs. Huxley's approval, and being myself quite in the mood for a holiday, I volunteered to wrestle with him, and succeeded, holding out as an inducement a visit to the volcanic region of the Auvergne with Scrope's classical volume, which we both knew and admired, as a guide book.

We started on July 2nd, I loaded with injunctions from his physician as to what his patient was to eat, drink, and avoid, how much he was to sleep and rest, how little to talk and walk, etc., that would have made the expedition a perpetual burthen to me had I not believed that I knew enough of my friend's disposition and ailments to be convinced that not only health but happiness would be our companions throughout. Sure enough, for the first few days, including a short stay in Paris, his spirits were low indeed, but this gave me the opportunity of appreciating his remarkable command over himself and his ever-present consideration for his companion. Not a word or gesture of irritation ever escaped him; he exerted himself to obey the instructions laid down; nay, more, he was instant in his endeavour to save me trouble at hotels, railway stations, and ticket offices. Still, some mental recreation was required to expedite recovery, and he found it first by picking up at a bookstall, a *History of the Miracles of Lourdes*, which were then exciting the religious fervour of France, and the interest of her scientific public. He entered with enthusiasm into the subject, getting together all the treatises upon it, favourable or the reverse, that were accessible, and I need hardly add, soon arrived at the conclusion, that the so-called miracles were in part illusions and for the rest delusions. As it may interest some of your readers to know what his opinion was in this the early stage of the manifestations, I will give it as he gave it to me. It was a case of two peasant children sent in the hottest

month of the year into a hot valley to collect sticks for
firewood washed up by a stream, when one of them after
stooping down opposite a heat-reverberating rock, was,
in rising, attacked with a transient vertigo, under which
she saw a figure in white against the rock. This bare
fact being reported to the curé of the village, all the
rest followed.

Soon after our arrival at Clermont Ferrand, your
father had so far recovered his wonted elasticity of spirits
that he took a keen interest in everything around, the
museums, the cathedral, where he enjoyed the conclusion
of the service by a military band which gave selections
from the Figlia del Regimento, but above all he appreciated the walks and drives to the geological features of
the environs. He reluctantly refrained from ascending
the Puy de Dome, but managed the Pic Parion, Gergovia,
Royat, and other points of interest without fatigue. . . .

After Clermont they visited the other four great
volcanic areas explored by Scrope, Mont Dore, the
Cantal, Le Puy, and the valley of the Ardèche.
Under the care of his friend, and relieved from the
strain of work, my father's health rapidly improved.
He felt no bad effects from a night at Mont Dore,
when, owing to the crowd of invalids in the little
town, no better accommodation could be found than
a couple of planks in a cupboard. Next day they
took up their quarters in an unpretentious cabaret
at La Tour d'Auvergne, one of the villages on the
slopes of the mountain, a few miles away.

Here (writes Sir J. Hooker), and for some time afterwards, on our further travels, we had many interesting
and amusing experiences of rural life in the wilder parts
of central France, its poverty, penury, and too often its

inconceivable impositions and overcharges to foreigners, quite consistently with good feeling, politeness, and readiness to assist in many ways.

By the 10th of July, nine days after setting out, I felt satisfied (he continues) that your father was equal to an excursion upon which he had set his heart, to the top of the Pic de Sancy, 4000 feet above La Tour and 7 miles distant.

It was on this occasion that the friends made what they thought a new discovery, namely evidence of glacial action in central France. Besides striated stones in the fields or built into the walls, they noticed the glaciated appearance of one of the valleys descending from the peak, and especially some isolated gigantic masses of rock on an open part of the valley, several miles away, as to which they debated whether they were low buildings or transported blocks. Sir Joseph visited them next day, and found they were the latter, brought down from the upper part of the peak.[1]

Le Puy offered a special attraction apart from scenery and geology. In the museum was the skeleton of a pre-historic man that had been found in the breccia of the neighbourhood, associated with the remains of the rhinoceros, elephant, and other extinct mammals. My father's sketch-book contains drawings of these bones and of the ravine where they were discovered, although in spite of directions from

---

[1] He published an account of these blocks in *Nature*, xiii. 31, 166, but subsequently found that glaciation had been observed by von Lassaul in 1872 and by Sir William Guise in 1870.

M. Aymard, the curator, he could not find the exact spot. Under the sketch is a description of the remains, in which he notes, " The bones do not look fresher than some of those of Elephas and Rhinoceros in the same or adjacent cases."

As for the final stage of the excursion :—

After leaving the Ardèche (continues Sir J. Hooker), with no Scrope to lead or follow, our scientific ardours collapsed. We had vague views as to future travel. Whatever one proposed was unhesitatingly acceded to by the other. A more happy-go-lucky pair of idlers never joined company.

As will be seen from the following letters, they made their way to the Black Forest, where they stayed till Sir Joseph's duties called him back to England, and my mother came out to join my father for the rest of his holiday.[1]

[1] You ask me (Sir Joseph adds) whether your father smoked on the occasion of this tour. Yes, he did, cigars in moderation. But the history of his addiction to tobacco that grew upon him later in life, dates from an earlier excursion that we took together, and I was the initiator of the practice. It happened in this wise ; he had been suffering from what was supposed to be gastric irritation, and, being otherwise "run down," we agreed to go, in company with Sir John Lubbock, on a tour to visit the great monoliths of Brittany. This was in 1867. On arriving at Dinan he suffered so much, that I recommended his trying a few cigarettes which I had with me. They acted as a charm, and this led to cigars, and finally, about 1875 I think, to the pipe. That he subsequently carried the use of tobacco to excess is, I think, unquestionable. I repeatedly remonstrated with him, at last I think (by backing his medical adviser) with effect.

I have never blamed myself for the "teaching him " to smoke, for the practice habitually palliated his distressing symptoms when nothing else did, nor can his chronic illness be attributed to the abuse of tobacco.

The following letters to Sir H. Roscoe and Dr. Tyndall were written during this tour :—

LE PUY, HAUTE LOIRE, FRANCE,
July 17, 1873.

MY DEAR ROSCOE—Your very kind letter reached me just as I was in the hurry of getting away from England, and I have been carrying it about in my pocket ever since.

Hooker and I have been having a charming time of it among the volcanoes of the Auvergne, and we are now on our way to those of the Velay and Vivarrais. The weather has been almost perfect. Perhaps a few degrees of temperature could have been spared now and then, especially at Clermont, of which somebody once said that having stayed there the climate of hell would have no terrors for him.

It has been warm in the Mont Dore country and in the Cantal, as it is here, but we are very high up, and there is a charming freshness and purity about the air.

I do not expect to be back before the end of September, and my lectures begin somewhere in the second week of October. After they commence I shall not be able to leave London even for a day, but I shall be very glad to come to the inauguration of your new buildings if the ceremony falls within my possible time. And you know I am always glad to be your guest.

I am thriving wonderfully. Indeed all that plagues me now is my conscience, for idling about when I feel full of vigour. But I promised to be obedient, and I am behaving better than Auld Clootie did when he fell sick.

I hope you are routing out the gout. This would be the p ace for you—any quantity of mineral waters.

Pray remember me very kindly to Mrs. Roscoe, and believe me, ever yours very faithfully,

T. H. HUXLEY.

HOTEL DE FRANCE, BADEN-BADEN,
July 30, 1873.

MY DEAR TYNDALL—We find ourselves here after a
very successful cruise in the Auvergne and Ardèche,
successful at least so far as beauty and geological interest
go. The heat was killing, and obliged us to give up all
notion of going to Ursines, as we had at first intended
to do. So we turned our faces north and made for
Grenoble, hoping for a breath of cool air from the
mountains of Dauphiny. But Grenoble was hotter even
than Clermont (which, by the way, quite deserves its
reputation as a competitor with hell), a neighbour's
drains were adrift close to the hotel, and we got poisoned
before we could escape. Luckily we got off with nothing
worse than a day or two's diarrhœa. After this the best
thing seemed to be to rush northward to Gernsbach,
which had been described to me as a sort of earthly
paradise. We reached the place last Saturday night,
and found ourselves in a big rambling hotel, crammed
full of people, and planted in the bottom of a narrow
valley, all hot and steaming. A large pigstye "convenient" to the house mingled its vapours with those of
the seventy or eighty people who eat and drank without
any other earthly occupation that we could discern
during the three days we were bound, by stress of
letters and dirty linen, to stop. On Monday we made
an excursion over here, prospecting, and the air was so
fresh and good, and things in general looked so promising
that I made up my mind to put up in Baden-Baden until
the wife joins me. She writes me that you talk of
leaving England on Friday, and I may remark that
Baden is on the high road to Switzerland. *Verbum sap.*

I am wonderfully better, and really feel ashamed of
loafing about when I might very well be at work. But
I have promised to make holiday, and make holiday
I will.

No proof of your answer to Forbes' biographer reached me before I left, so I suppose you had not received one in time. I am dying to see it out.

Hooker is down below, but I take upon myself to send his love. He is in great force now that he has got rid of his Grenoble mulligrubs.—Ever yours,

T. H. HUXLEY.

After parting company with Hooker, he paid a flying visit to Professor Bonnet at Geneva ; then he was joined by his wife and son for the last three weeks of the holiday, which were spent at Baden and in the Bernese Oberland. Before this, he writes home :—

> I feel quite a different man from what I was two months ago, and you will say that you have a much more creditable husband than the broken-down old fellow who has been a heart-ache to you so long, when you see me. The sooner you can get away the better. If the rest only does you as much good as it does me, I shall be very happy.

AXENSTEIN, LUZERNE,
Aug. 24, 1873.

MY DEAR TYNDALL—The copies of your booklet[1] intended for Hooker and me reached me just as I left Baden last Tuesday. Hooker had left me for home a fortnight before, and I hardly know whether to send his to Kew or keep them for him till I return. I have read mine twice, and I think that nothing could be better than the tone you have adopted. I did not suspect that you had such a shot in your locker as the answer to Forbes about the direction of the "crevasses" referred to by Rendu. It is a deadly thrust ; and I shall be curious to see what sort of parry the other side will attempt.

---

[1] " Principal Forbes and his Biographers."

For of course they will attempt something. Scotland is,
I believe, the only country in the world in which you
can bring an action for "putting to silence" an adversary
who will go on with an obviously hopeless suit. The
lawgivers knew the genius of the people ; and it is to be
regretted that they could not establish a process of the
same sort in scientific matters.

I wrote to you a month ago to tell you how we had
been getting on in France. Hooker and I were very
jolly, notwithstanding the heat, and I think that the
Vivarrais is the most instructive country in the world for
seeing what water can do in cutting down the hardest
rocks. Scrope's book is very good on the whole, though
the pictures are a little overdone.

My wife and Leonard met me at Cologne on the 11th.
Then we went on to Baden and rested till last Tuesday,
when we journeyed to Luzerne and, getting out of that
hot and unsavoury hole as fast as we could, came here
last Thursday.

We find ourselves very well off. The hotel is perched
up 1800 feet above the lake, with a beautiful view of
Pilatus on the west and of the Urner See on the south.
On the north we have the Schwyz valley, so that we are
not shut in, and the air is very good and fresh. There
are plenty of long walks to be had without much fatigue,
which suits the wife. Leonard promises to have very
good legs of his own with plenty of staying power. I
have given him one or two sharp walks, and I find he
has plenty of vigour and endurance. But he is not
thirteen yet and I do not mean to let him do overmuch,
though we are bent on a visit to a glacier. I began to
tell him something about the glaciers the other day, but
I was promptly shut up with, "Oh yes! I know all
about that. It's in Dr. Tyndall's book"—which said
book he seems to me to have got by heart. He is the
sweetest little fellow imaginable ; and either he has
developed immensely in the course of the last year, or I

have never been so much thrown together with him alone, and have not had the opportunity of making him out.

You are a fatherly old bachelor, and will not think me a particularly great donkey for prattling on in this way about my swan, who probably to unprejudiced eyes has a power of goose about him.

I suppose you know that in company with yourself and Hooker, the paternal gander (T. H. H.) has been honoured by the King of Sweden and made into a Polar Goose by the order of the North Star. Hooker has explained to the Swedish Ambassador that English officials are prohibited by order in Council from accepting foreign orders, and I believe keeps the cross and ribbon on these conditions. If it were an ordinary decoration I should decline with thanks, but I am told it is a purely scientific and literary affair like the Prussian "pour le mérite"; so when I get back I shall follow Hooker's line.

I met Laugel on board the Luzerne steamboat the other day, and he told me that you were at the Belalp—gallivanting as usual, and likely to remain there for some time. So I send this on the chance of finding you.—With best love from us all, ever yours, T. H. HUXLEY.

I am as well as I ever was in my life—regularly set up—in token whereof I have shaved off my beard.

In another letter to his wife, dated August 8, from Baden, there is a very interesting passage about himself and his aims. He has just been speaking about his son's doings at school :—

I have been having a great deal of talk with myself about my future career too, and I have often thought over what you say in the letter you wrote to the Puy. I don't quite understand what —— meant about the disputed reputation, unless it is a reputation for getting into disputes. But to say truth I am not greatly concerned about any reputation except that of being entirely honest and straightforward, and that reputation I think and hope I have.

For the rest . . . the part I have to play is not to found a new school of thought or to reconcile the antagonisms of the old schools. We are in the midst of a gigantic movement greater than that which preceded and produced the Reformation, and really only the continuation of that movement. But there is nothing new in the ideas which lie at the bottom of the movement, nor is any reconcilement possible between free thought and traditional authority. One or other will have to succumb after a struggle of unknown duration, which will have as side issues vast political and social troubles. I have no more doubt that free thought will win in the long run than I have that I sit here writing to you, or that this free thought will organise itself into a coherent system, embracing human life and the world as one harmonious whole. But this organisation will be the work of generations of men, and those who further it most will be those who teach men to rest in no lie, and to rest in no verbal delusions. I may be able to help a little in this direction—perhaps I may have helped already. For the present, however, I am disposed to draw myself back entirely into my own branch of physical science. There is enough and to spare for me to do in that line, and, for years to come, I do not mean to be tempted out of it.

Strangely enough, this was the one thing he was destined not to do. Official work multiplied about him. From 1870 to 1884 only two years passed without his serving on one or two Royal Commissions. He was Secretary of the Royal Society from 1871 to 1880, and President from 1883 to his retirement, owing to ill-health, in 1885. He became Dean as well as Professor of Biology in the College of Science, and Inspector of Fisheries. Though he still managed to find some time for anatomical investigations, and would steal a precious hour or half-hour by driving back from the Home Office to his laboratory at South Kensington before returning home to St. John's Wood, the amount of such work as he was able to publish could not be very great.

His most important contributions during this decennium (writes Sir M. Foster) were in part continuations of his former labours, such as the paper and subsequent full memoir on Stagonolepis, which appeared in 1875 and 1877, and papers on the Skull. The facts that he called a communication to the Royal Society, in 1875,[1] on Amphioxus, a preliminary note, and that a paper read to the Zoological Society in 1876, on Ceratodus Forsteri, was marked No. 1 of the series of Contributions to Morphology, showed that he still had before him the prospect of much anatomical work, to be accomplished when opportunity offered; but, alas! the opportunity which came was small, the preliminary note had no full successor, and No. 1 was only followed, and that after an interval of seven years, by a brief No. 2. A paper "On the Characters of the Pelvis," in the *Proceedings of the Royal Society*, in 1879, is full of suggestive thought, but its concluding passages seem to suggest that others, and not he himself, were to carry out the ideas. Most of the papers of this decennium deal with vertebrate morphology, and are more or less connected with his former researches, but in one respect, at least, he broke quite fresh ground. He had chosen the crayfish as one of the lessons for the class in general biology spoken of above, and was thus drawn into an interesting study of crayfishes, by which he

---

[1] Written 1874.

was led to a novel and important analysis of the gill plumes as evidence of affinity and separation. He embodied the main results of his studies in a paper to the Zoological Society, and treated the whole subject in a more popular style in a book on the Crayfish. In a somewhat similar way, having taken the dog as an object lesson in mammalian anatomy for his students, he was led to a closer study of that common animal, resulting in papers on that subject to the Zoological Society in 1880, and in two lectures at the Royal Institution in 1880. He had intended so to develop this study of the dog as to make it tell the tale of mammalian morphology; but this purpose, too, remained unaccomplished.

Moreover, though he sent one paper (on Hyperodapedon Gordoni) to the Geological Society as late as 1887, yet the complete breakdown of his health in 1885, which released him from nearly all his official duties, at the same time dulled his ardour for anatomical pursuits. Stooping over his work became an impossibility.

Though he carried about him, as does every man of like calibre and experience, a heavy load of fragments of inquiry begun but never finished, and as heavy a load of ideas for promising investigations never so much as even touched, though his love of science and belief in it might never have wavered, though he never doubted the value of the results which further research would surely bring him, there was something working within him which made his hand, when turned to anatomical science, so heavy that he could not lift it. Not even that which was so strong within him, the duty of fulfilling a promise, could bring him to the work. In his room at South Kensington, where for a quarter of a century he had laboured with such brilliant effect, there lay on his working table for months, indeed for years, partly dissected specimens of the rare and little studied marine animal, Spirula, of which he had promised to contribute an account to the Reports of the "Challenger" Expedition, and hard by lay the already engraven plates; there was still wanted nothing more than some further investigation and the working out of the results. But it seemed as if some hidden hands were always being stretched out to keep him from the task; and eventually another labourer had to complete it. (*Ibid.*)

The remaining letters of this year include several to Dr. Dohrn, which show the continued interest my father took in the great project of the Biological Station at Naples, which was carried through in spite of many difficulties. He had various books and proceedings of learned societies sent out at Dr. Dohrn's request (I omit the details), and proposed a scheme for raising funds towards completing the building when the contractor failed. The scheme, however, was not put into execution.

4 MARLBOROUGH PLACE,
Feb. 24, 1873.

MY DEAR DOHRN—I was very glad to receive the fine sealed letter, and to get some news of you—though to be sure there is not much of you in the letter, but all is "Station, Station."

I congratulate you heartily on your success with your undertaking, and I only wish I could see England represented among the applicants for tables. But you see England is so poor, and the present price of coals obliges her to economise.

I envy you your visit from "Pater Anchises" Baer, and rejoice to hear that the grand old man is well and strong enough to entertain such a project. I wish I could see my way to doing the like. I have had a long bout of illness—ever since August—but I am now very much better, indeed, I hope I may say quite well. The weariness of all this has been complicated by the trouble of getting into a new house, and in addition a lawsuit brought by a knavish neighbour, in the hope of extracting money out of me.

I am happy to say, however, that he has just been thoroughly and effectually defeated. It has been a new experience for me, and I hope it may be my last as well as my first acquaintance with English law, which is a luxury of the most expensive character.

If Dr. Kleinenberg is with you, please to tell him, with my compliments and thanks for the copy of his Memoir, that I went over his Hydra paper pretty carefully in the summer, and satisfied myself as to the correctness of his statements about the structure of the ectoderm and about the longitudinal fibres. About the Endoderm I am not so clear, and I often found indications of delicate circular fibres in close apposition with the longitudinal ones. However, I had not time to work all this out, and perhaps might as well say nothing about it.

Pray make my very kind remembrances to Mr. Grant. I trust that his dramas may have a brilliant reception.

The Happy Family flourishes. But we shall look to your coming to see us. The house is big enough now to give you a bedroom, and you know you will have no lack of a welcome.

I have said nothing about my wife (who has been in a state not only of superhuman, but of superfeminine, activity for the last three months) meaning to leave her the last page to speak for herself.

With best compliments to the "ladies downstairs," ever yours very faithfully, T. H. HUXLEY.

4 MARLBOROUGH PLACE,
Oct. 17, 1873.

MY DEAR DOHRN—Your letter reached me nearly a week ago, and I have been turning over its contents in my mind as well as I could, but have been able to come to no clear conclusion until now. I have been incessantly occupied with other things.

I will do for you, and gladly, anything I would do for myself, but I could not apply on my own behalf to any of those rich countrymen of mine, unless they were personally well known to me, and I had the opportunity of feeling my way with them. But if you are disposed to apply to any of the people you mention, I shall be only too glad to back your application with all the force I am master of. You may make use of my name to any extent as guarantor of the scientific value and importance of your undertaking and refer any one to whom you may apply to me. It may be, in fact, that this is all you want, but as you have taken to the caprice of writing in my tongue instead of in that vernacular, idiomatic and characteristically Dohrnian German in which I delight, I am not so sure about your meaning. There is a rub for you. If you write to me in English again I will send the letter back without paying the postage.

In any case let me have a precise statement of your financial position. I may have a chance of talking to some Croesus, and the first question he is sure to ask me is—How am I to know that this is a stable affair, and that I am not throwing my money into the sea? . . .

(Referring to an unpleasant step it seemed necessary to take) . . . you must make up your mind to act decidedly and take the consequences. No good is ever done in this world by hesitation. . . .

I hope you are physically better. Look sharply after your diet, take exercise and defy the blue-devils, and you will weather the storm.—Ever yours very faithfully,

T. H. HUXLEY.

Tyndall, who had not attended the 1873 meeting of the British Association, had heard that some local opposition had been offered to his election as President for the Belfast meeting in 1874, and had written :—

I wish to heaven you had not persuaded me to accept that Belfast duty. They do not want me. . . . But Spottiswoode assures me that no individual offered the slightest support to the two unscientific persons who showed opposition.

The following was written in reply :—

4 MARLBOROUGH PLACE,
Sept. 25, 1873.

MY DEAR TYNDALL—I am sure you are mistaken about the Belfast people. That blundering idiot of—— wanted to make himself important and get up a sort of " Home Rule " agitation in the Association, but nobody backed him and he collapsed. I am at your disposition for whatever you want me to do, as you know, and I am sure Hooker is of the same mind. We shall not be ashamed when we meet our enemies in the gate.

The grace of God cannot entirely have deserted you since you are aware of the temperature of that ferocious epistle. Reeks,[1] whom I saw yesterday, was luxuriating in it, and said (confound his impudence) that it was quite my style. I forgot to tell him, by the bye, that I had resigned in your favour ever since the famous letter to Carpenter. Well, so long as you are better after it there is no great harm done.

Somebody has sent me the two numbers of Scribner with Blauvelt's articles on " Modern Skepticism." They seem to be very well done, and he has a better appreciation of the toughness of the job before him than any of

---

[1] The late Trenham Reeks, Registrar of the School of Mines, and Curator of the Museum of Practical Geology.

the writers of his school with whom I have met.   But it
is rather cool of you to talk of his pitching into Spencer
when you are chief target yourself.   I come in only *par
parenthèse*, and I am glad to see that people are beginning
to understand my real position, and to separate me from
such raging infidels as you and Spencer.—Ever thine,

T. H. HUXLEY.

He was unable to attend the opening of Owens
College this autumn, and having received but a scanty
account of the proceedings, wrote as follows :—

4 MARLBOROUGH PLACE, LONDON, N.W.,
Oct. 16, 1873.

My DEAR ROSCOE—I consider myself badly used.   Nobody has sent me a Manchester paper with the proceedings
of the day of inauguration, when, I hear, great speeches
were made.

I *did* get *two* papers containing your opening lecture,
and the "Fragment of a Morality," for which I am duly
grateful, but two copies of one day's proceedings are not
the same thing as one copy of two days' proceedings, and
I consider it is very disrespectful to a Governor (large G)
not to let him know what went on.

By all accounts which have reached me it was a great
success, and I congratulate you heartily.   I only wish
that I could have been there to see.—Ever yours very
faithfully,   T. H. HUXLEY.

The autumn brought a slow improvement in
health—

I am travelling (he writes) between the two stations of
dyspepsia and health thus (illustrated by a zigzag with
"mean line ascending").

The sympathy of the convalescent appears in
various letters to friends who were ill.   Thus, in reply

to Mr. Hyde Clarke, the philologist and, like himself,
a member of the Ethnological Society, he writes :—

(Nov. 18, 1873)—I am glad to learn two things from
your note—first, that you are getting better ; second, that
there is hope of some good coming out of that Ashantee
row, if only in the shape of rare vocables.

My attention is quite turned away from Anthropological
matters at present, but I will bear your question in mind
if opportunity offers.

A letter to Professor Rolleston at Oxford gives a
lively account of his own ailments, which could only
have been written by one now recovering from them,
while the illness of another friend raised a delicate
point of honour, which he laid before the judgment
of Mr. Darwin, more especially as the latter had been
primarily concerned in the case.

4 MARLBOROUGH PLACE,
Oct. 16, 1873.

MY DEAR ROLLESTON—A note which came from Mrs.
Rolleston to my wife the other day, kindly answering
some inquiries of ours about the Oxford Middle Class
Examination, gave us but a poor account of your health.

This kind of thing won't do, you know. Here is
—— ill, and I doing all I can to persuade him to go
away and take care of himself, and now comes ill news of
you.

Is it dyspeps again ? If so follow in my steps. I
mean to go about the country, with somebody who can
lecture, as the "horrid example"—cured. Nothing but
gross and disgusting intemperance, Sir, was the cause of
all my evil. And now that I have been a teetotaller for
nine months, and have cut down my food supply to
about half of what I used to eat, the enemy is beaten.

I have carried my own permissive bill, and no canteen (except for my friends who still sit in darkness) is allowed on the premises. And as this is the third letter I have written before breakfast (a thing I never could achieve in the days when I wallowed in the stye of Epicurus), you perceive that I am as vigorous as ever I was in my life.

Let me have news of you, and believe me—Ever yours very faithfully,   T. H. H.

                                        ATHENÆUM CLUB,
                                          Nov. 3, 1873.

MR DEAR DARWIN—You will have heard (in fact I think I mentioned the matter when I paid you my pleasant visit the other day) that —— is ill and obliged to go away for six months to a warm climate. It is a great grief to me, as he is a man for whom I have great esteem and affection, apart from his high scientific merits, and his symptoms are such as cause very grave anxiety. I shall be happily disappointed if that accursed consumption has not got hold of him.

The college authorities have behaved as well as they possibly could to him, and I do not suppose that his enforced retirement for a while gives him the least pecuniary anxiety, as his people are all well off, and he himself has an income apart from his college pay. Nevertheless, under such circumstances, a man with half a dozen children always wants all the money he can lay hands on; and whether he does or no, he ought not to be allowed to deprive himself of any, which leads me to the gist of my letter. His name was on your list as one of those hearty friends who came to my rescue last year, and it was the only name which made me a little uneasy, for I doubted whether it was right for a man with his responsibilities to make sacrifices of this sort. However, I stifled that feeling, not seeing what else I could do without wounding him. But now my conscience won't let me be, and I do not think that any consideration ought to deter me from getting his contribution back to him somehow or other There is no one to whose judgment on a point of honour I would defer more readily than yours, and I am quite sure you will agree with me. I really am quite unhappy and ashamed to think of myself as vigorous and well at the expense of his denying himself any rich man's caprice he might take a fancy to.

So, my dear, good friend, let me know what his contribution was, that I may get it back to him somehow or other, even if I go like Nicodemus privily and by night to his bankers.—Ever yours faithfully,

T. H. H.

CHAPTER V

1874

My father's health continued fairly good in 1874, and while careful to avoid excessive strain he was able to undertake nearly as much as before his illness outside his regular work at South Kensington, the Royal Society, and on the Royal Commission. To this year belong three important essays, educational and philosophical. From February 25 to March 3 he was at Aberdeen, staying first with Professor Bain, afterwards with Mr. Webster, in fulfilment of his first duty as Lord Rector[1] to deliver an address to the students. Taking as his subject "Universities, Actual and Ideal," he then proceeded to vindicate, historically and philosophically, the claims of natural science to take the place from which it had so long been ousted in the universal culture which a University professes to give. More especially he demanded an improved system of education in the

---

[1] It may be noted that between 1860 and 1890 he and Professor Bain were the only Lord Rectors of Aberdeen University elected on non-political grounds.

medical school, a point to which he gave practical effect in the Council of the University.

In an ideal University, as I conceive it, a man should be able to obtain instruction in all forms of knowledge, and discipline in the use of all the methods by which knowledge is obtained. In such a University the force of living example should fire the student with a noble ambition to emulate the learning of learned men, and to follow in the footsteps of the explorers of new fields of knowledge. And the very air he breathes should be charged with that enthusiasm for truth, that fanaticism of veracity, which is a greater possession than much learning ; a nobler gift than the power of increasing knowledge ; by so much greater and nobler than these, as the moral nature of man is greater than the intellectual ; for veracity is the heart of morality. (*Coll. Ess.* iii. 189, *sqq.*)

As for the "so-called ' conflict of studies,' " he exclaims—

One might as well inquire which of the terms of a Rule of Three sum one ought to know in order to get a trustworthy result. Practical life is such a sum, in which your duty multiplied into your capacity and divided by your circumstances gives you the fourth term in the proportion, which is your deserts, with great accuracy.

The knowledge on which medical practice should be based is "the sort of practical, familiar, finger-end knowledge which a watchmaker has of a watch," the knowledge gained in the dissecting-room and laboratory.

Until each of the greater truths of anatomy and physiology has became an organic part of your minds—until you would know them if you were roused and questioned in the middle of the night, as a man knows the geography of his native place and the daily life of his home. That is the sort of knowledge which, once obtained, is a lifelong possession. Other occupations may fill your minds—it may grow dim and seem to be forgotten—but there it is, like the inscription on a battered and defaced coin, which comes out when you warm it.

Hence the necessity to concentrate the attention on these cardinal truths, and to discard a number of extraneous subjects commonly supposed to be requisite whether for general culture of the medical student or to enable him to correct the possible mistakes of druggists. Against this "Latin fetish" in medical education, as he used to call it, he carried on a lifelong campaign, as may be gathered from his published essays on medical education, and from letters given in later chapters of this book. But there is another side to such limitation in professional training. Though literature is an essential in the preliminary, general education, culture is not solely dependent upon classics.

Moreover, I would urge that a thorough study of Human Physiology is in itself an education broader and more comprehensive than much that passes under that name. There is no side of the intellect which it does not call into play, no region of human knowledge into which either its roots or its branches do not extend ; like the Atlantic between the Old and the New Worlds, its waves wash the shores of the two worlds of matter and of mind; its tributary streams flow from both; through its waters, as yet unfurrowed by the keel of any Columbus, lies the road, if such there be, from the one to the other; far away from that North-west Passage of mere speculation, in which so many brave souls have been hopelessly frozen up.

Of the address he writes to his wife, February 27 :—

I have just come back from the hall in which the address was delivered, somewhat tired. The hall was very large, and contained, I suppose, a couple of thousand people, and the students made a terrific row at intervals, though they were quiet enough at times. As the address took me an hour and a half to deliver, and my voice has been very shaky ever since I have been here, I did not dare to put too much strain upon it, and I suspect that the people at the end of the hall could have heard very little. However, on the whole, it went off better than I expected.

And to Professor Baynes :—

I am very glad you liked my address. The students were abnormally quiet for the first half-hour, and then made up for their reticence by a regular charivari for the rest of the time. However, I was consoled by hearing that they were much quieter than usual.

Dr. John Muir's appreciation is worth having. It did not occur to me that what I had to say would interest people out of Britain, but to my surprise I had an application from a German for permission to translate the address the other day.

Again to his wife, March 1 :—

. . . I was considerably tired after my screed on Friday, but Bain and I took a long walk, and I was

fresh again by dinner-time. I dined with the Senators
at a hotel in the town, and of course had to make a
speech or two. However I cut all that as fast as I
could. They were all very apologetic for the row the
students made. After the dinner one of the Professors
came to ask me if I would have any objection to attend
service in the College Chapel on Sunday, as the students
would like it. I said I was quite ready to do anything
it was customary for the Rector to do, and so this
morning in half an hour's time I shall be enduring the
pains and penalties of a Presbyterian service.

There was to have been another meeting of the
University Court yesterday, but the Principal was
suffering so much from an affection of the lungs that
I adjourned the meeting till to-morrow. Did I tell
you that I carried all my resolutions about improving
the medical curriculum? Fact, though greatly to my
astonishment. To-morrow we go in for some reforms in
the arts curriculum, and I expect that the job will be
tougher.

I send you a couple of papers—*Scotsman*, with a very
good leading article, and the *Aberdeen Herald* also with
a leading article, which is as much favourable as was to
be expected. . . . The Websters are making me promise
to bring you and one of the children here next autumn.
They are wonderfully kind people.

March 2.—My work here finishes to-day. There is
a meeting of the Council at one o'clock, and before that
I am to go and look over laboratories and collections
with sundry Professors. Then there is the supper at
half-past eight and the inevitable speeches, for which I
am not in the least inclined at present. I went officially
to the College Chapel yesterday, and went through a
Presbyterian service for the first time in my life. May
it be the last!

Then to lunch at Professor Struthers' and back here
for a small dinner-party. I am standing it all well, for

the weather is villanous and there is no getting any exercise. I shall leave here by the twelve o'clock train to-morrow.

On August 2 he delivered an address on "Joseph Priestley" (*Coll. Ess.* iii. 1) at Birmingham, on the occasion of the presentation of a statue of Priestley to that town. The biography of this pioneer of science and of political reform, who was persecuted for opinions that have in less than a century become commonplaces of orthodox thought, suggested a comparison between those times and this, and evoked a sincere if not very enthusiastic tribute to one who had laboured to better the world, not for the sake of worldly honour, but for the sake of truth and right.

As the way to Birmingham lay through Oxford, he was asked by Professor Ray Lankester, then a Fellow of Exeter College, if he could not break his journey there, and inspect the results of his investigations on Lymnæus. The answer was as follows:—

We go to Birmingham on Friday by the three o'clock train, but there is no chance of stopping at Oxford either going or coming, so that unless you bring a Lymnæus or two (under guise of periwinkles for refreshment) to the carriage door I shall not be able to see them.

The following letters refer both to this address on Priestley, and to the third of the important addresses of this year, that "On the Hypothesis that Animals are Automata, and its History" (*Coll. Ess.* i. 199, see also p. 131 below) The latter was delivered at Belfast before the British Association under Tyndall's presidency. It appears that only a month before, he had not so much as decided upon his subject—indeed, was thinking of something quite different.

The first allusion in these letters is to a concluding phase of Tyndall's controversy upon the claims of the late Principal Forbes in the matter of Glacier theory :—

4 MARLBOROUGH PLACE, LONDON, N.W.,
June 24, 1874.

MY DEAR TYNDALL—I quite agree with your Scotch friend in his estimate of Forbes, and if he were alive and the controversy beginning, I should say draw your picture in your best sepia or lampblack. But I have been thinking over this matter a good deal since I received your letter, and my verdict is, leave that tempting piece of portraiture alone.

The world is neither wise nor just, but it makes up for all its folly and injustice by being damnably sentimental, and the more severely true your portrait might be the more loud would be the outcry against it. I should say publish a new edition of your *Glaciers of the Alps*, make a clear historical statement of all the facts showing Forbes's relations to Rendu and Agassiz, and leave the matter to the judgment of your contemporaries. That will sink in and remain when all the hurly-burly is over.

I wonder if that address is begun, and if you are going to be as wise and prudent as I was at Liverpool. When I think of the temptation I resisted on that occasion, like Clive when he was charged with peculation, "I marvel at my own forbearance!" Let my example be a burning and a shining light to you. I declare I have horrid misgivings of your kicking over the traces.

The "x" comes off on Saturday next, so let your ears burn, for we shall be talking about you. I have just begun my lectures to Schoolmasters, and I wish they were over, though I am very well on the whole.

Griffith [1] wrote to ask for the title of my lecture at Belfast, and I had to tell him I did not know yet. I shall not begin to think of it till the middle of July when these lectures are over.

The wife would send her love, but she has gone to Kew to one of Hooker's receptions, taking Miss Jewsbury,[2] who is staying with us. I was to have gone to the College of Physicians' dinner to-night, but I was so weary when I got home that I made up my mind to send an excuse. And then came the thought that I had not written to you.—Ever yours sincerely,

T. H. HUXLEY.

The next letter is in reply to Tyndall, who had written as follows from Switzerland on July 15 :—

I confess to you that I am far more anxious about your condition than about my own ; for I fear that after your London labour the labour of this lecture will press heavily upon you. I wish to Heaven it could be transferred to other shoulders.

I wish I could get rid of the uncomfortable idea that I have drawn upon you at a time when your friend and brother ought to be anxious to spare you every labour. . . .

P.S.—Have just seen the Swiss *Times;* am intensely disgusted to find that while I was brooding over the calamities possibly consequent on your lending me a

---

[1] For many years secretary to the British Association.

[2] Miss Geraldine Jewsbury (1812-80) the novelist, and friend of the Carlyles. After 1866 she lived at Sevenoaks.

hand, that you have been at the Derby Statue, and are to make an oration apropos of the Priestley Statue in Birmingham on the 1st August! ! !

4 MARLBOROUGH PLACE, LONDON, N.W.,
July 22, 1874.

MY DEAR TYNDALL—I hope you have been taking more care of your instep than you did of your leg in old times. Don't try mortifying the flesh again.

I was uncommonly amused at your disgustful wind-up after writing me such a compassionate letter. I am as jolly as a sandboy so long as I live on a minimum and drink no alcohol, and as vigorous as ever I was in my life. But a late dinner wakes up my demoniac colon and gives me a fit of blue devils with physical precision.

Don't believe that I am at all the places in which the newspapers put me. For example, I was not at the Lord Mayor's dinner last night. As for Lord Derby's statue, I wanted to get a lesson in the art of statue unveiling. I help to pay Dizzie's salary, so I don't see why I should not get a wrinkle from that artful dodger.

I plead guilty to having accepted the Birmingham invitation.[1] I thought they deserved to be encouraged for having asked a man of science to do the job instead of some noble swell ; and, moreover, Satan whispered that it would be a good opportunity for a little ventilation of wickedness. I cannot say, however, that I can work myself up into much enthusiasm for the dry old Unitarian who did not go very deep into anything. But I think I may make him a good peg whereon to hang a discourse on the tendencies of modern thought.

I was not at the Cambridge pow-wow—not out of prudence, but because I was not asked. I suppose that decent respect towards a secretary of the Royal Society was not strong enough to outweigh University objections to the

---

[1] To unveil the statue of Joseph Priestley. See above, p. 127

incumbent of that office. It is well for me that I expect
nothing from Oxford or Cambridge, having burned my
ships so far as they were concerned long ago.

I sent your note on to Knowles as soon as it arrived,
but I have heard nothing from him. I wrote to him
again to-night to say that he had better let me see it in
proof if he is going to print it. I am right glad you
find anything worth reading again in my old papers. I
stand by the view I took of the origin of species now as
much as ever.

Shall I not see the address? It is tantalising to hear
of your progress and not to know what is in it.

I am thinking of taking Development for the subject
of my evening lecture,[1] the concrete facts made out in
the last thirty years without reference to Evolution. If
people see that it is Evolution, that is Nature's fault, and
not mine.

We are all flourishing, and send our love.—Ever
yours faithfully,   T. H. HUXLEY.

The paper on Animal Automatism is in effect an
enlargement of a short paper read before the Metaphysical Society in 1871, under the title of "Has a
Frog a Soul?" It begins with a vindication of Descartes as a great physiologist, doing for the physiology
of motion and sensation that which Harvey had done
for the circulation of the blood. A series of propositions which constitute the foundation and essence of
the modern physiology of the nervous system are
fully expressed and illustrated in the writings of
Descartes. Modern physiological research, which
has shown that many apparently purposive acts are

[1] *I.e.* at the British Association ; he actually took "Animals
as Automata."

performed by animals, and even by men, deprived of consciousness, and therefore of volition, is at least compatible with the theory of automatism in animals, although the doctrine of continuity forbids the belief that "such complex phenomena as those of consciousness first make their appearance in man." And if the volitions of animals do not enter into the chain of causation of their actions at all, the fact lays at rest the question, "How is it possible to imagine that volition, which is a state of consciousness, and, as such, has not the slightest community of nature with matter in motion, can act upon the moving matter of which the body is composed, as it is assumed to do in voluntary acts?"

As for man, the argumentation, if sound, holds equally good. States of consciousness are immediately caused by molecular changes of the brain-substance, and our mental conditions are simply the symbols in consciousness of the changes which take place automatically in the organism.

As for the bugbear of the "logical consequences" of this conviction, "I may be permitted to remark (he says), that logical consequences are the scarecrows of fools and the beacons of wise men." And if St. Augustine, Calvin, and Jonathan Edwards have held in substance the view that men are conscious automata, to hold this view does not constitute a man a fatalist, a materialist, nor an atheist. And he takes occasion once more to declare that he ranks among none of these philosophers.

Not among fatalists, for I take the conception of necessity to have a logical, and not a physical foundation; not among materialists, for I am utterly incapable of conceiving the existence of matter if there is no mind in which to picture that existence; not among atheists, for the problem of the ultimate cause of existence is one which seems to me to be hopelessly out of reach of my poor powers. Of all the senseless babble I have ever had occasion to read, the demonstrations of these philosophers who undertake to tell us all about the nature of God would be the worst, if they were not surpassed by the still greater absurdities of the philosophers who try to prove that there is no God.

This essay was delivered as an evening address on August 24, the Monday of the Association week. A vast stir had been created by the treatment of deep reaching problems in Professor Tyndall's presidential address; interest was still further excited by this unexpected excursion into metaphysics. "I remember," writes Sir M. Foster, "having a talk with him about the lecture before he gave it. I think I went to his lodgings—and he sketched out what he was going to say. The question was whether, in view of the Tyndall row, it was wise in him to take the line he had marked out. In the end I remember his saying, 'Grasp your nettle, that is what I have got to do.'" But apart from the subject, the manner of the address struck the audience as a wonderful *tour de force*. The man who at first disliked public speaking, and always expected to break down on the platform, now, without note or reference of any kind, discoursed for an hour and a half upon a complex and difficult subject, in the very words which he had thought out and afterwards published.

This would have been a remarkable achievement if he had planned to do so and had learned up his speech; but the fact was that he was compelled to speak offhand on the spur of the moment. He describes the situation in a letter of February 6, 1894, to Professor Ray Lankester:—

I knew that I was treading on very dangerous ground, so I wrote out uncommonly full and careful notes, and had them in my hand when I stepped on to the platform.

Then I suddenly became aware of the bigness of the audience, and the conviction came upon me that, if I looked at my notes, not one half would hear me. It was a bad ten seconds, but I made my election and turned the notes face downwards on the desk.

To this day, I do not exactly know how the thing managed to roll itself out; but it did, as you say, for the best part of an hour and a half.

There's a story *pour vous encourager* if you are ever in a like fix.

He writes home on August 20 :—

Johnny's address went off exceedingly well last night. There was a mighty gathering in the Ulster Hall, and he delivered his speech very well. The meeting promises to be a good one, as there are over 1800 members already, and I daresay they will mount up to 2000 before the end. The Hookers' arrangements[1] all went to smash as I rather expected they would, but I have a very good clean lodging well outside the town where I can be quiet

---

[1] *I.e.* for the members of the *x*-club and their wives to club together at Belfast.

if I like, and on the whole I think that is better, as I shall be able to work up my lectures in peace. . . .

August 21.—Everything is going on very well here. The weather is delightful, and under these circumstances my lodgings here with John Ball for a companion turns out to be a most excellent arrangement. I need not say that I was speaking more or less all day long. *Ça va sans dire*, though, by the way, that is a bull induced by the locality. I am not going on any of the excursions on Sunday. I am going to have a quiet day here when everybody will suppose that I have accepted everybody else's invitation to be somewhere else. The Ulster Hall, in which the addresses are delivered, seems to me to be a terrible room to speak in, and I mean to nurse my energies all Monday. I sent you a cutting from one of the papers containing an account of me that will amuse you. The writer is evidently disappointed that I am not a turbulent savage.

August 25 :—

. . . My work is over and I start for Kingstown, where I mean to sleep to-night, in an hour. I have just sent you a full and excellent report of my lecture.[1] I am glad to say it was a complete success. I never was in better voice in my life, and I spoke for an hour and a half without notes, the people listening as still as mice. There has been a great row about Tyndall's address, and I had some reason to expect that I should have to meet a frantically warlike audience. But it was quite other-wise, and though I spoke my mind with very great plainness, I never had a warmer reception. And I am not without hope that I have done something to allay the storm, though, as you may be sure, I did not sacrifice plain speaking to that end. . . . I have been most creditably quiet here, and have gone to no dinners or

---

[1] "On Animals as Automata" : see above.

breakfasts or other such fandangoes except those I accepted
before leaving home.   Sunday I spent quietly here,
thinking over my lecture and putting my peroration,
which required a good deal of care, into shape.   I
wandered out into the fields in the afternoon, and sat a
long time thinking of all that had happened since I was
here a young beginner, two and twenty, and . . . you
were largely in my thoughts, which were full of blessings
and tender memories.

I had a good night's work last night.   I dined with
the President of the College, then gave my lecture.   After
that I smoked a bit with Foster till eleven o'clock, and
then I went to the *Northern Whig* office to see that the
report of my lecture was all right.   It is the best paper
here, and the Editor had begged me to see to the report,
and I was anxious myself that I should be rightly represented.   So I sat there till a quarter past one having the
report read and correcting it when necessary.   Then I
came home and got to bed about two.   I have just been
to the section and read my paper there to a large
audience who cannot have understood ten words of it, but
who looked highly edified, and now I have done.   Our
lodging has turned out admirably, and Ball's company
has been very pleasant.   So that the fiasco of our arrangements was all for the best.

I take the account of this last-mentioned paper in
Section D from the report in *Nature* :—

Professor Huxley opened the last day of the session
with an account of his recent observations on the development of the *Columella auris* in Amphibia.   (He described
it as an outgrowth of the periotic capsule, and therefore
unconnected with any visceral arch.) . . .

In the absence of Mr. Parker there was no one
competent to criticise the paper from personal knowledge ;
but a word dropped as to the many changes in the

accepted homologies of the ossicula auditus, elicited a
masterly and characteristic exposition of the series of
new facts, and the modifications of the theory they have
led to, from Reichert's first observations down to the
present time. The embryonic structures grew and
shaped themselves on the board, and shifted their
relations in accordance with the views of successive
observers, until a graphic epitome of the progress of
knowledge on the subject was completed.

He and Parker indeed (to whom he signs himself, "Ever yours amphibially") had been busy, not
only throughout 1874, but for several years earlier,
examining the development of the *Amphibia*, with a
particular view to the whole theory of the vertebrate
skull, for which he had done similar work in 1857
and 1858. Thus in May 4, 1870, he writes to
Parker :—

I read all the most important part of your Frog-
paper last night, and a grand piece of work it is—more
important, I think, in all its bearings than anything
you have done yet.

From which premisses I am going to draw a conclusion
which you do not expect, namely, that the paper must
by no manner of means go into the Royal Society in its
present shape. And for the reasons following :—

In the first place, the style is ultra-Parkerian. From a
literary point of view, my dear friend, you remind me
of nothing so much as a dog going home. He has a goal
before him which he will certainly reach sooner or later, but
first he is on this side the road, and now on that; anon,
he stops to scratch at an ancient rat-hole, or maybe he
catches sight of another dog, a quarter of a mile behind,
and bolts off to have a friendly, or inimical sniff. In
fact, his course is . . . (here a tangled maze is drawn)

not ——. In the second place, you must begin with an earlier stage. . . . That is the logical starting-point of the whole affair.

Will you come and dine at 6 on Saturday, and talk over the whole business ?

If you have drawings of earlier stages you might bring them. I suspect that what is wanted might be supplied in plenty of time to get the paper in.

In 1874 he re-dissects the skull of *Axolotl* to clear up the question as to the existence of the "ventral head or pedicle" which Parker failed to observe: "If you disbelieve in that pedicle again, I shall be guilty of an act of personal violence." Later, "I am benevolent to all the world, being possessed of a dozen live axolotls and four or five big dead mesobranchs. Moreover, I am going to get endless Frogs and Toads by judicious exchange with Gunther.[1] We will work up the Amphibia as they have not been done since they were crea—I mean evolved."

The question of the pedicle comes up again when he simplifies some of Parker's results as to the development of the *Columella auris* in the Frog. "Your suprahyomandibular is nothing but the pedicle of the suspensorium over again. It has nothing whatever to do with the columella auris. . . . The whole thing will come out as simply as possible without any of your coalescences and combotherations. How you will hate me and the pedicle."

Tracing the development of the *columella* was a

[1] Dr. A. C. L. G. Gunther, of the British Museum, where he was appointed Keeper of the Department of Zoology in 1875.

long business, but it grew clearer as young frogs of various ages were examined. "Don't be aggravated with yourself," he writes to Parker in July, "it's tough work, this here Frog." And on August 5: "I have worked over Toad and I have worked over Frog, and I tell an obstinate man that s.h.m. (suprahyomandibular) is a figment—or a vessel, whichever said obstinate man pleases." The same letter contains what he calls his final views on the *columella*, but by the end of the year he has gone further, and writes :—

Be prepared to bust-up with all the envy of which your malignant nature is capable. The problem of the vertebrate skull is solved. Fourteen segments or thereabouts in *Amphioxus;* all but one (barring possibilities about the ear capsule) aborted in higher vertebrata. Skull and brain of *Amphioxus* shut up like an opera-hat in higher vertebrata. So ! (Sketch in illustration.)

P.S.—I am sure you will understand the whole affair from this. Probably published it already in *Nature* !

A letter to the *Times* of July 8, 1874, on women's education, was evoked by the following circumstances. Miss Jex Blake's difficulties in obtaining a medical education have already been referred to (p. 95). A further discouragement was her rejection at the Edinburgh examination. Her papers, however, were referred to Huxley, who decided that certain answers were not up to the standard.

As Miss Jex Blake may possibly think that my decision was influenced by prejudice against her cause, allow me to add that such prejudice as I labour under lies in the opposite direction. Without seeing any reason to believe that women are, on the average, so strong physically, intellectually, or morally, as men, I cannot shut my eyes to the fact that many women are much better endowed in all these respects than many men, and I am at a loss to understand on what grounds of justice or public policy a career which is open to the weakest and most foolish of the male sex should be forcibly closed to women of vigour and capacity.

We have heard a great deal lately about the physical disabilities of women. Some of these alleged impediments, no doubt, are really inherent in their organisation, but nine-tenths of them are artificial—the products of their modes of life. I believe that nothing would tend so effectually to get rid of these creations of idleness, weariness, and that "over-stimulation of the emotions" which, in plainer-spoken days, used to be called wantonness, than a fair share of healthy work, directed towards a definite object, combined with an equally fair share of healthy play, during the years of adolescence; and those who are best acquainted with the acquirements of an average medical practitioner will find it hardest to believe that the attempt to reach that standard is like to prove exhausting to an ordinarily intelligent and well-educated young woman.

The Marine Biological Station at Naples was still struggling for existence, and to my father's interest in it is due the following letter, one of several to Dr. Dohrn, whose marriage took place this summer :—

4 MARLBOROUGH PLACE,
June 24, 1874.

MY DEAR DOHRN—Are you married yet or are you not? It is very awkward to congratulate a man upon

what may not have happened to him, but I shall assume
that you are a benedict, and send my own and my wife's
and all the happy family's good wishes accordingly.   May
you have as good a wife and as much a "happy family"
as I have, though I would advise you—the hardness of
the times being considered—to be satisfied with fewer
than seven members thereof.

I hear excellent accounts of the progress of the Station
from Lankester, and I hope that it is now set on its legs
permanently.   As for the English contribution, you must
look upon it simply as the expression of the hearty
goodwill of your many friends in the land of fogs, and of
our strong feeling that where you had sacrificed so much
for the cause of science, we were, as a matter of duty,—
quite apart from goodwill to you personally—bound to
do what we could, each according to his ability.

Darwin is, in all things, noble and generous—one of
those people who think it a privilege to let him help.   I
know he was very pleased with what you said to him.
He is working away at a new edition of the *Descent of
Man*, for which I have given him some notes on the
brain question.

And apropos of that, how is your own particular
brain ?   I back la belle M—— against all the physicians
in the world—even against mine own particular Æscu-
lapius, Dr. Clark—to find the sovereignest remedy against
the blue devils.

Let me hear from you—most abominable of correspondents as I am.   And why don't you send Madame's
photograph that you have promised ?—Ever yours very
faithfully,

T. H. HUXLEY.

Pray give my kind remembrances to your father.

4 MARLBOROUGH PLACE,
March 31, 1874.

MY DEAR DARWIN—The brain business [1] is more than half done, and I will soon polish it off and send it to you. We are going down to Folkestone for a week on Thursday, and I shall take it with me.

I do not know what is doing about Dohrn's business at present. Foster took it in hand, but the last time I heard he was waiting for reports from Dew and Balfour.

You have been very generous as always; and I hope that other folk may follow your example, but like yourself I am not sanguine.

I have had an *awfully* tempting offer to go to Yankee-land on a lecturing expedition, and I am seriously thinking of making an experiment next spring.

The chance of clearing two or three thousand pounds in as many months is not to be sneezed at by a *père de famille.* I am getting sick of the state of things here.—Ever yours faithfully,   T. H. HUXLEY.

I have heard no more about the spirit. photographs!

4 MARLBOROUGH PLACE,
April 16, 1874.

MY DEAR DARWIN—Put my contribution into the smallest type possible, for it will be read by none but anatomists; and never mind where it goes.

I am glad you agree with me about the hand and foot and skull question. As Ward [2] said of Mill's opinions, you can only account for the views of Messrs. —— and Co. on the supposition of "grave personal sin" on their part.

I had a letter from Dohrn a day or two ago in which

---

[1] A note on the brain in man and the apes for the second edition of the *Descent of Man.*

[2] W. G. Ward. (See i. 454.)

he tells me he has written to you.   I suspect he has been very ill.

Let us know when you are in town, and believe me—
Ever yours very faithfully,   T. H. HUXLEY.

The allusion in the letter of March 31 to certain "spirit photographs" refers to a series of these wonderful productions sent to him by a connection of Mr. Darwin's, who was interested in these matters, and to whom he replied, showing how the effect might have been produced by simple mechanical means.

It was at this gentleman's house that in January a carefully organised seance was held, at which my father was present incognito, so far as the medium was concerned, and on which he wrote the following report to Mr. Darwin, referred to in his *Life*, vol. iii. p. 187.

It must be noted that he had had fairly extensive experience of spiritualism; he had made regular experiments with Mrs. Haydon at his brother George's house (the paper on which these are recorded is undated, but it must have been before 1863); he was referred to as a disbeliever in an article in the *Pall Mall Gazette* during January 1869, as a sequel to which a correspondent sent him an account of the confessions of the Fox girls, who had started spiritualism forty years before. At the houses of other friends, he had attended seances and met mediums, by whom he was most unfavourably impressed.

Moreover, when invited to join a committee of

investigation into spiritualistic manifestations, he replied :—

I regret that I am unable to accept the invitation of the Committee of the Dialectical Society to co-operate with a committee for the investigation of "Spiritualism"; and for two reasons. In the first place, I have not time for such an inquiry, which would involve much trouble and (unless it were unlike all inquiries of that kind I have known) much annoyance. In the second place, I take no interest in the subject. The only case of "Spiritualism" I have had the opportunity of examining into for myself, was as gross an imposture as ever came under my notice. But supposing the phenomena to be genuine—they do not interest me. If anybody would endow me with the faculty of listening to the chatter of old women and curates in the nearest cathedral town, I should decline the privilege, having better things to do. And if the folk in the spiritual world do not talk more wisely and sensibly than their friends report them to do, I put them in the same category. The only good that I can see in the demonstration of the truth of "Spiritualism" is to furnish an additional argument against suicide. Better live a crossing-sweeper than die and be made to talk twaddle by a "medium" hired at a guinea a *seance*.[1]

To the report above mentioned, Prof. G. Darwin, who also was present, added one or two notes and corrections.

REPORT ON SEANCE

Jan. 27, 1874.

We met in a small room at the top of the house with a window capable of being completely darkened by a

---

[1] Quoted from a review in the *Daily News*, October 17, 1871, of the Report on Spiritualism of the Committee of the London Dialectical Society.

shutter and curtains opposite the door. A small light table with two flaps and four legs, unsteady and easily moved, occupied the middle of the room, leaving not much more than enough space for the chairs at the sides. There was a chair at each end, two chairs on the fireplace side, and one on the other. Mr. X (the medium) was seated in the chair at the door end, Mr. Y (the host) in the opposite chair, Mr. G. Darwin on the medium's right, Mr. Huxley on his left, Mr. Z between Mr. Huxley and Mr. [Darwin] Y. The table was small enough to allow these five people to rest their hands on it, linking them together. On the table was a guitar which lay obliquely across it, an accordion on the medium's side of the guitar, a couple of paper horns, a Japanese fan, a matchbox, and a candlestick with a candle.

At first the room was slightly darkened (leaving plenty of light from the window, however) and we all sat round for half an hour. My right foot was against the medium's left foot, and two fingers of my right hand had a good grip of the little finger of his left hand. I compared my hand (which is *not* small and *is* strong) with his, and was edified by its much greater massiveness and strength. (No, we didn't link until the darkness. G. D.)

G. D.'s left hand was, as I learn, linked with medium's right hand, and left foot on medium's [left] right foot.

We sat thus for half an hour as aforesaid and nothing happened.

The room was next thoroughly darkened by shutting the shutters and drawing the curtains. Nevertheless, by great good fortune I espied three points of light, coming from the lighted passage outside the door. One of these came beneath the door straight to my eye, the other two were on the wall (or on a press) obliquely opposite. By still greater good fortune, these three points of light had such a position in reference to my eye that they gave me three straight lines traversing and bounding the space in which the medium sat, and I at once saw that if

Medium moved his body forwards or backwards he must occult one of my three rays. While therefore taking care to feel his foot and keep a good grip of his hand, I fixed my eyes intently on rays A and B. For I felt sure that I could trust to G. D. keeping a sharp look-out on the right hand and foot; and so no instrument of motion was left to the medium but his body and head, the movements of which could not have been discernible in absolute darkness. Nothing happened for some time. At length a very well executed muscular twitching of the arm on my side began, and I amused myself by comparing it with the convulsions of a galvanised frog's leg, but at the same time kept a very bright look-out on my two rays A and B.

The twitchings ceased, and then after a little time A was shut out. B then became obscure, and A became visible. " Ho ho ! " thought I, " Medium's head is well over the table. Now we are going to have some manifestations." Immediately followed a noise obviously produced by the tumbling over of the accordion and some shifting of the position of the guitar. Next came a twanging—very slight, but of course very audible—of some of the strings, during which B was invisible By and by B and A became visible again, and Medium's voice likewise showed that he had got back to his first position. But after he had returned to this position there was a noise of the guitar and other things on the table being stirred, and creeping noises like something light moving over the table. But no more actual twanging.

To my great disgust, G. D. now began to remark that he saw two spots of light, which I suppose must have had the same origin as my rays A and B, and, moreover, that something occasionally occulted one or other of them. (Note: No, not till we changed places. G. H. D.) I blessed him for spoiling my game, but the effect was excellent. Nothing more happened. By and by, after

some talk about these points of light, the medium suggested that this light was distracting, and that we had
better shut it out. The suggestion was very dexterously
and indirectly made, and was caught up more strongly
(I think by Mr. Z). Anyhow, we agreed to stop out all
light. The circle was broken, and the candle was lighted
for this purpose. I then took occasion to observe that
the guitar was turned round into the position noted in
the margin, the end being near my left hand. On
examining it I found a longish end of one of the catgut
strings loose, and I found that by sweeping this end over
the strings I could make quite as good twangs as we
heard. I could have done this just as well with my
mouth as with my hand—and I could have pulled the
guitar about by the end of the catgut in my mouth and
so have disturbed the other things—as they were disturbed.

Before the candle was lighted some discussion arose as
to why the spirits would not do any better (started by
Mr. Y and Mr. Z, I think), in which the medium joined.
It appeared that (in the opinion of the spirits as interpreted by the medium) we were not quite rightly placed.
When the discussion arose I made a bet with myself that
the result would be that either I or G. D. would have to
change places with somebody else. And I won my wager
(I have just paid it with the remarkably good cigar I am
now smoking). G. D. had to come round to my side,
Mr. Z went to the end, and Mr. Y took G. D.'s place.
"Good, Medium," said I to myself. "Now we shall see
something." We were in pitch darkness, and all I could
do was to bring my sense of touch to bear with extreme
tension upon the medium's hand—still well in my grip.

Before long Medium became a good deal convulsed at
intervals, and soon a dragging sound was heard, and Mr.
Y told us that the arm-chair (mark its position) had
moved up against his leg, and was shoving against him.
By degrees the arm-chair became importunate, and by

the manner of Mr. Y's remarks it was clear that his attention was entirely given to its movements.

Then I felt the fingers of the medium's left hand become tense—in such a manner as to show that the muscles of the left arm were contracting sympathetically with those of the other arm, on which a considerable strain was evidently being put. Mr. Y's observations upon the eccentricities of the arm-chair became louder— a noise was heard as of the chair descending on the table and shoving the guitar before it (while at the same time, or just before, there was a crash of a falling thermometer), and the tension of the left arm ceased. The chair had got on to the table. Says the medium to Mr. Y, "Your hand was against mine all the time." " Well, no," replied Mr. Y, "not quite. For a moment as the chair was coming up I don't think it was." But it was agreed that this momentary separation made no difference. I said nothing, but, like the parrot, thought the more. After this nothing further happened. But conversation went on, and more than once the medium was careful to point out that the chair came upon the table while his hand was really in contact with Mr. Y's.

G. D. will tell you if this is a fair statement of the facts. I believe it is, for my attention was on the stretch for those mortal two hours and a half, and I did not allow myself to be distracted from the main points in any way. My conclusion is that Mr. X is a cheat and an impostor, and I have no more doubt that he got Mr. Y to sit on his right hand, knowing from the turn of his conversation that it would be easy to distract his attention, and that he then moved the chair against Mr. Y with his leg, and finally coolly lifted (it) on to the table, than that I am writing these lines.   T. H. H.

As Mr. G. Darwin wrote of the seance, " It has given me a lesson with respect to the worthlessness of evidence which I shall always remember, and besides will make me very diffident in trusting myself. Unless I had seen it, I could not have believed in the evidence of any one with such perfect *bona fides* as Mr. Y being so worthless."

On receiving this report Mr. Darwin wrote (*Life*, ii. p. 188) :—

Though the seance did tire you so much, it was, I think, really worth the exertion, as the same sort of things are done at all the seances . . . . and now to my mind an enormous weight of evidence would be requisite to make me believe in anything beyond mere trickery.

The following letter to Mr. Morley, then editor of the *Fortnightly Review*, shows that my father was already thinking of writing upon Hume, though he did not carry out this intention till 1878.

The article referred to in the second letter is that on Animals as Automata.

4 MARLBOROUGH PLACE, N.W.,
June 4, 1874.

MY DEAR MR. MORLEY—I assure you that it was a great disappointment to me not to be able to visit you, but we had an engagement of some standing for Oxford.

Hume is frightfully tempting—I thought so only the other day when I saw the new edition advertised—and now I would gladly write about him in the *Fortnightly* if I were only sure of being able to keep any engagement to that effect I might make.

But I have yet a course of lectures before me, and an evening discourse to deliver at the British Association— to say nothing of opening the Manchester Medical School in October—and polishing off a lot of scientific work. So you see I have not a chance of writing about Hume for months to come, and you had much better not trust to such a very questionable reed as I am.—Ever yours very faithfully,   T. H. HUXLEY.

4 MARLBOROUGH PLACE, N.W.,
Nov. 15, 1874.

MY DEAR MORLEY—Many thanks for your abundantly
sufficient cheque—rather too much, I think, for an article
which had been gutted by the newspapers.

I am always very glad to have anything of mine in
the *Fortnightly*, as it is sure to be in good company;
but I am becoming as spoiled as a maiden with many
wooers.   However, as far as the *Fortnightly* which is my
old love, and the *Contemporary* which is my new, are
concerned, I hope to remain as constant as a persistent
bigamist can be said to be.

It will give me great pleasure to dine with you, and
Dec. 1 will suit me excellently well.—Ever yours very
faithfully,   T. H. HUXLEY.

The year winds up with a New Year's greeting to
Professor Haeckel.

4 MARLBOROUGH PLACE, LONDON, N.W.,
Dec. 28, 1874.

MY DEAR HAECKEL—This must reach you in time to
wish you and yours a happy New Year in English
fashion.   May your shadow never be less, and may
all your enemies, unbelieving dogs who resist the Prophet
of Evolution, be defiled by the sitting of jackasses upon
their grandmothers' graves ! an oriental wish appropriate
to an ex-traveller in Egypt.

I have written a notice of the " Anthropogenie" for
the Academy, but I am so busy that I am afraid I should
never have done it—but for being put into a great
passion—by an article in the *Quarterly Review* for last
July, which I read only a few days ago.   My friend
Mr. ——, to whom I had to administer a gentle punishment some time ago, has been at the same tricks again,
but much worse than his former performance—you will
see that I have dealt with him as you deal with a

"Pfaffe." [1] There are "halb-Pfaffen" as well as "halb-Affen." [2] So if what I say about "Anthropogenie" seems very little—to what I say about the *Quarterly Review*—do not be offended. It will all serve the good cause.

I have been working very hard lately at the lower vertebrata, and getting out results which will interest you greatly. Your suggestion that Rathke's canals in *Amphioxus* [3] are the Wolffian ducts was a capital shot, but it just missed the mark because Rathke's canals do not exist. Nevertheless there are two half canals, the dorsal walls of which meet in the raphe described by Stieda, and the plaited lining of this wall (*a*) is, I believe, the renal organ. Moreover, I have found the skull and brain of *Amphioxus*, both of which are very large (like a vertebrate embryo's) instead of being rudimentary as we all have thought, and exhibit the primitive segmentation of the "Urwirbelthier" [4] skull.

Thus the skull of *Petromyzon* answers to about fourteen segments of the body of *Amphioxus*, fused together and indistinguishable in even the earliest embryonic state of the higher vertebrata.

Does this take your breath away? Well, in due time you shall be convinced. I sent in a brief notice to the last meeting of the Royal Society, which will soon be in your hands.

I need not tell you of the importance of all this. It is unlucky for Semper that he has just put *Amphioxus* out of the *Vertebrata* altogether—because it is demonstrable that *Amphioxus* is nearer than could have been hoped to the condition of the primitive vertebrate—a far more regular and respectable sort of ancestor than even you suspected. For you see "Acrania" will have to go.

I think we must have an English translation of the *Anthropogenie*. There is great interest in these questions

---

[1] Parson.   [2] Lit. half-apes ; the Prosimiæ and Lemurs.   [3] The Lancelet.   [4] Primitive vertebrate.

now, and your book is very readable, to say nothing of its higher qualities.

My wife (who sends her kindest greetings) and I were charmed with the photograph. [As for our] publication in that direction, the seven volumes are growing into stately folios. You would not know them.—Ever yours very faithfully,   T. H. HUXLEY.

How will you read this scrawl now that Gegenbaur is gone ?

In the article here referred to, a review of a book by Prof. G. H. Darwin, a personal attack of an unjustifiable character was made upon him, and through him, upon Charles Darwin. The authorship of the review in question had come to be known, and Huxley writes to his friend :—

I entirely sympathise with your feeling about the attack on George. If anybody tries that on with my boy L., the old wolf will show all the fangs he has left by that time, depend upon it. . . .

You ought to be like one of the blessed gods of Elysium, and let the inferior deities do battle with the infernal powers. Moreover, the severest and most effectual punishment for this sort of moral assassination is quietly to ignore the offender and give him the cold shoulder. He knows why he gets it, and society comes to know why, and though society is more or less of a dunderhead, it has honourable instincts, and the man in the cold finds no cloak that will cover him.

CHAPTER VI

1875–1876

In the year 1875 the bitter agitation directed against experimental physiology came to a head. It had existed in England for several years. In 1870, when President of the British Association, Huxley had been violently attacked for speaking in defence of Brown Sequard, the French physiologist. The name of vivisection, indifferently applied to all experiments on animals, whether carried out by the use of the knife or not, had, as Dr. (afterwards Sir) William Smith put it, the opposite effect on many minds to that of the "blessed word Mesopotamia." Misrepresentation was rife even among the most estimable and well-meaning of the opponents of vivisection, because they fancied they saw traces of the practice everywhere, all the more, perhaps, for not having sufficient technical knowledge for proper discrimination. One of the most flagrant instances of this kind of thing was a letter in the *Record* charging Huxley with advocating vivisections before children, if not by them. Passages from the Introduction to his

Elementary Physiology, urging that beginners should be shown the structures under discussion, examples for which could easily be provided from the domestic animals, were put side by side with later passages in the book, such, for instance, as statements of fact as to the behaviour of severed nerves under irritation. A sinister inference was drawn from this combination, and published as fact without further verification. Of this he remarks emphatically in his address on "Elementary Instruction in Physiology," 1877 (*Collected Essays*, iii. 300):

It is, I hope, unnecessary for me to give a formal contradiction to the silly fiction, which is assiduously circulated by the fanatics who not only ought to know, but do know, that their assertions are untrue, that I have advocated the introduction of that experimental discipline which is absolutely indispensable to the professed physiologist, into elementary teaching.

Moreover, during the debates on the Vivisection Bill in 1876, the late Lord Shaftesbury made use of this story. Huxley was extremely indignant, and wrote home :—

Did you see Lord Shaftesbury's speech in Tuesday's *Times*? I saw it by chance,[1] and have written a sharp letter to the *Times*.

This letter appeared on May 26, when he wrote again :—

You will have had my note, and know all about Lord Shaftesbury and his lies by this time. Surely you could

---

[1] Being in Edinburgh, he had been reading the Scotch papers, and "the reports of the Scotch papers as to what takes place in Parliament are meagre."

not imagine on any authority that I was such an idiot as to recommend boys and girls to perform experiments which are difficult to skilled anatomists, to say nothing of other reasons.

LETTER TO THE *TIMES*

In your account of the late debate in the House of Lords on the Vivisection Bill, Lord Shaftesbury is reported to have said that in my *Lessons in Elementary Physiology*, it is strongly insisted that such experiments as those subjoined shall not merely be studied in the manual, but actually repeated, either by the boys and girls themselves or else by the teachers in their presence, as plainly appears from the preface to the second edition.

I beg leave to give the most emphatic and unqualified contradiction to this assertion, for which there is not a shadow of justification either in the preface to the second edition of my *Lessons* or in anything I have ever said or written elsewhere. The most important paragraph of the preface which is the subject of Lord Shaftesbury's misquotation and misrepresentation stands as follows:—

"For the purpose of acquiring a practical, though elementary, acquaintance with physiological anatomy and histology, the organs and tissues of the commonest domestic animals afford ample materials. The principal points in the structure and mechanism of the heart, the lungs, the kidneys, or the eye of man may be perfectly illustrated by the corresponding parts of a sheep; while the phenomena of the circulation, and many of the most important properties of living tissues are better shown by the common frog than by any of the higher animals."

If Lord Shaftesbury had the slightest theoretical or practical acquaintance with the subject about which he is so anxious to legislate, he would know that physiological anatomy is not exactly the same thing as experimental physiology; and he would be aware that the recommendations of the paragraph I have quoted might be fully carried into effect without the performance of even a solitary "vivisection." The assertion that I have ever suggested or desired the introduction of vivisection into the teaching of elementary physiology in schools is, I repeat, contrary to fact.

On the next day (May 27) appeared a reply from Lord Shaftesbury, in which his entire good faith is equally conspicuous with his misapprehension of the subject.

LORD SHAFTESBURY'S REPLY

The letter from Professor Huxley in the *Times* of this morning demands an immediate reply.

The object that I supposed the learned professor had in view was gathered from the prefaces to the several editions of his work on *Elementary Physiology*.

The preface to the first edition states that "the following lessons in elementary physiology are, primarily, intended to serve the purpose of a text-book for teachers and learners in boys' and girls' schools."

It was published, therefore, as a manual for the young, as well as the old.

Now, any reader of the preface to the first edition would have come to the conclusion that teachers and learners could acquire something solid, and worth having, from the text-book before them. But the preface to the second edition nearly destroys that expectation. Here is the passage :—"It will be well for those who attempt to study elementary physiology to bear in mind the important truth that the knowledge of science which is attainable by mere reading, though infinitely better than ignorance, is knowledge of a very different kind from that which arises from direct contact with fact."

"Direct contact with fact!" What can that mean (so, at least, very many ask) but a declaration, on high authority, to teachers and learners that vivisection alone can give them any real and effective instruction?

But the subsequent passage is still stronger, for it states "that the worth of the pursuit of science, as an intellectual discipline, is almost lost by those who only seek it in books."

Is not language like this calculated to touch the zeal and vanity of teachers and learners at the very quick, and urge them to improve their own minds and stand well in the eyes of the profession and the public by positive progress in experimental physiology? Ordinary readers, most people would think, could come to no other conclusion.

But a disclaimer from Professor Huxley is enough; I am sorry to have misunderstood him; and I must ask his pardon. I sincerely rejoice to have received such an assurance that his great name shall never be used for such a project as that which excited our fears.

On this he wrote :—

You will have seen Lord Shaftesbury's reply to my letter. I thought it frank and straightforward, and I have written a private letter [1] to the old boy of a placable and proper character.

In 1874 he had also had a small passage of arms with the late Mr. W. E. Forster, then Vice-President of the Council, upon the same subject. Mr. Forster was about to leave office, and when he gave his official authorisation for summer courses of lectures

[1] "Huxley, the Professor, has written me a very civil, nay kind, letter. I replied in the same spirit." (Lord Shaftesbury, *Life and Work*, iii. 373, June 3, 1876.)

at South Kensington on Biology, Chemistry, Geology, etc., he did so with the special proviso that there be no vivisection experiments in any of the courses, and further, appended a Memorandum, explaining the reasons on which he acted.

Now, although Huxley was mentioned by name as having taken care to avoid inflicting pain in certain previous experiments which had come to Mr. Forster's knowledge, the memorandum evoked from him a strong protest to the Lord President, to whom, as Mr. Forster expressly intimated, an appeal might properly be made.

To begin with, the memorandum contained a mistake in fact, referring to his regular course at South Kensington experiments which had taken place two years before at one of the Courses to Teachers. This course was non-official; Huxley's position in it was simply that of a private person to whom the Department offered a contract, subject to official control and criticism, so far as touched that course, and entirely apart from his regular position at the School of Mines. The experiments of 1872 were performed, as he had reason to believe, with the full sanction of the Department. If the Board chose to go back upon what had happened two years before, he was of course subject to their criticism, but then he ought in justice to be allowed to explain in what these experiments really consisted. What they were appears from a note to Sir J. Donnelly :—

My dear Donnelly—It will be the best course, perhaps, if I set down in writing what I have to say respecting the vivisections for physiological purposes which have been performed here, and concerning which you made me a communication from the Vice-President of the Council this morning.

I have always felt it my duty to defend those physiologists who, like Brown Sequard, by making experiments on living animals, have added immensely not only to scientific physiology, but to the means of alleviating human suffering, against the often ignorant and sometimes malicious clamour which has been raised against them.

But personally, indeed I may say constitutionally, the performance of experiments upon living and conscious animals is extremely disagreeable to me, and I have never followed any line of investigation in which such experiments are required.

When the course of instruction in Physiology here was commenced, the question of giving experimental demonstrations became a matter of anxious consideration with me. It was clear that, without such demonstrations, the subject could not be properly taught. It was no less clear from what had happened to me when, as President of the British Association, I had defended Brown Sequard, that I might expect to meet with every description of abuse and misrepresentation if such demonstrations were given.

It did not appear to me, however, that the latter consideration ought to weigh with me, and I took such a course as I believe is defensible against everything but misrepresentation.

I gave strict instructions to the Demonstrators who assisted me that no such experiments were to be performed, unless the animal were previously rendered insensible to pain either by destruction of the brain or by the administration of anæsthetics, and I have every reason to believe that my instructions were carried out. I do not see what I can do beyond this, or how I can give Mr. Forster any better guarantee than is given in my assurance that my dislike to the infliction of pain both as a matter of principle and of feeling is quite as strong as his own can be.

If Mr. Forster is not satisfied with this assurance, and with its practical result that our experiments are made only on non-sentient animals, then I am afraid that my position as teacher of Physiology must come to an end.

If I am to act in that capacity I cannot consent to be prohibited from showing the circulation in a frog's foot because the frog is made slightly uncomfortable by being tied up for that purpose; nor from showing the fundamental properties of nerves, because extirpating the brain of the same animal inflicts one-thousandth part of the prolonged suffering which it undergoes when it makes its natural exit from the world by being slowly forced down the throat of a duck, and crushed and asphyxiated in that creature's stomach.

I shall be very glad to wait upon Mr. Forster if he desires to see me. Of course I am most anxious to meet his views as far as I can, consistently with my position as a person bound to teach properly any subject in which he undertakes to give instruction. But I am quite clear as to the amount of freedom of action which it is necessary I should retain, and if you will kindly communicate the contents of this letter to the Vice-President of the Council, he will be able to judge for himself how far his sense of what is right will leave me that freedom, or render it necessary for me to withdraw from what I should regard as a false position.

But there was a further and more vital question. He had already declared through Major (now Sir John) Donnelly, that he would only undertake a course which involved no vivisection. Further to require an official assurance that he would not do that which he had explicitly affirmed he did not intend to do, affected him personally, and he therefore declined the proposal made to him to give the course in question.

It followed from the fact that experiments on animals formed no part of his official course, and from his refusal under the circumstances to undertake the non-official course, that his opinions and present practices in regard to the question of vivisection did not come under their Lordships' jurisdiction, and he protested against the introduction of his name, and of the approbation or disapprobation of his views, into an official document relating to a matter with which he had nothing to do.

In an intermediate paragraph of the same document, he could not resist asking for an official definition of vivisection as forbidden, in its relation to the experiments he had made to the class of teachers.

I should have to ask whether it means that the teacher who has undertaken to perform no " vivisection experiments " is thereby debarred from inflicting pain, however slight, in order to observe the action of living matter; for it might be said to be unworthy quibbling, if, having accepted the conditions of the minute, he thought himself at liberty to inflict any amount of pain, so long as he did not actually cut.

But if such is the meaning officially attached to the word " vivisection," the teacher would be debarred from showing the circulation in a frog's foot or in a tadpole's tail; he must not show an animalcule, uncomfortably fixed under the microscope, nor prick his own finger for the sake of obtaining a drop of living blood. The living particles which float in that liquid undoubtedly feel as much (or as little) as a frog under the influence of anæsthetics, or deprived of its brain, does; and the teacher who shows his pupils the wonderful phenomena exhibited by dying blood, might be charged with gloating over the agonies of the colourless corpuscles, with quite as much justice as I have been charged with inciting boys and girls to cruelty by describing the results of physiological experiments, which they are as likely to attempt as they are to determine the longitude of their schoolroom.

However, I will not trouble your Lordship with any further indication of the difficulties which, as I imagine, will attend the attempt to carry the Minute into operation, if instruction is to be given in Physiology, or even in general Biology.

The upshot of the matter was that the Minute was altered so as to refer solely to future courses, and on February 20 he wrote to Mr. Forster:—

I cannot allow you to leave office without troubling you with the expression of my thanks for the very great kindness and consideration which I have received from you on all occasions, and particularly in regard to the question of vivisection, on which I ventured to some extent, though I think not very widely or really, to differ from you.

The modification which you were good enough to make in your minute removed all my objections to undertaking the Summer Course.

And I am sure that if that course had happened to be a physiological one I could do all I want to do in the way of experiment, without infringing the spirit of your minute, though I confess that the letter of it would cause me more perplexity.

As to his general attitude to the subject, it must be noted, as said above in the letter to Sir J. Donnelly, that he never followed any line of research involving experiments on living and conscious animals. Though, as will be seen from various letters, he considered such experiments justifiable, his personal feelings prevented him from performing them himself. Like Charles Darwin, he was very fond of animals, and our pets in London found in him an indulgent master.

But if he did not care to undertake such experiments personally, he held it false sentiment to blame others who did disagreeable work for the good of humanity, and false logic to allow pain to be inflicted in the cause of sport while forbidding it for the cause of science. (See his address on "Instruction in Elementary Physiology," *Coll. Essays*, iii. 300 *seq.*) Indeed, he declared that he trusted to the fox-hunting instincts of the House of Commons rather than to any real interest in science in that body, for a moderate treatment of the question of vivisection.

The subject is again dealt with in "The Progress of Science," 1887 (*Coll. Essays*, i. 122 *seq.*), from which I may quote two sentences :—

The history of all branches of science prove that they must attain a considerable stage of development before they yield practical "fruits '; and this is eminently true of physiology.

Unless the fanaticism of philozoic sentiment overpowers the voice of humanity, and the love of dogs and cats supersedes that of one's neighbour, the progress of experimental physiology and pathology will, indubitably, in course of time, place medicine and hygiene upon a rational basis.

The dangers of prohibition by law are discussed in a letter to Sir W. Harcourt :—

You wish me to say what, in my opinion, would be the effect of the total suppression of experiments on living animals on the progress of physiological science in this country.

I have no hesitation in replying that it would almost entirely arrest that progress.   Indeed, it is obvious that such an effect must follow the measure, for a man can no more develop a true conception of living action out of his inner consciousness than he can that of a camel Observation and experiment alone can give us a real foundation for any kind of Natural Knowledge, and any one who is acquainted with the history of science is aware that not a single one of all the great truths of modern physiology has been established otherwise than by experiment on living things.

Happily the abolition of physiological experiment in this country, should such a fatal legislative mistake ever be made, will be powerless to arrest the progress of science elsewhere.   But we shall import our physiology as we do our hock and our claret from Germany and France ; those of our young physiologists and pathologists who can afford to travel will carry on their researches in Paris and in Berlin, where they will be under no restraint whatever, or it may be that the foreign laboratories will carry out the investigations devised here by the few persons who have the courage, in spite of all obstacles, to attempt to save British science from extinction.

I doubt if such a result will contribute to the diminution of animal suffering.   I am sure that it will do as much harm as anything can do to the English school of Physiology, Pathology, and Pharmacology, and therefore to the progress of rational medicine.

Another letter on the subject may be given, which was written to a student at a theological college, in reply to a request for his opinion on vivisection, which was to be discussed at the college debating society.

GRAND HOTEL, EASTBOURNE,
Sept. 29, 1890.

DEAR SIR—I am of the opinion that the practice of performing experiments on living animals is not only reconcilable with true humanity, but under certain circumstances is imperatively demanded by it.

Experiments on living animals are of two kinds. First, those which are made upon animals which, although living, are incapable of sensation, in consequence of the destruction or the paralysis of the sentient machinery.

I am not aware that the propriety of performing experiments of this kind is seriously questioned, except in so far as they may involve some antecedent or subsequent suffering. Of course those who deny that under any circumstances it can be right to inflict suffering on other sentient beings for our own good, must object to even this much of what they call cruelty. And when they prove their sincerity by leaving off animal food ; by objecting to drive castrated horses, or indeed to employ animal labour at all ; and by refusing to destroy rats, mice, fleas, bugs and other sentient vermin, they may expect sensible people to listen to them, and sincere people to think them other than sentimental hypocrites.

As to experiments of the second kind, which do not admit of the paralysis of the sentient mechanism, and the performance of which involves severe prolonged suffering to the more sensitive among the higher animals, I should be sorry to make any sweeping assertion. I am aware of a strong personal dislike to them, which tends to warp my judgment, and I am prepared to make any allowance for those who, carried away by still more intense dislike, would utterly prohibit these experiments.

But it has been my duty to give prolonged and careful attention to this subject, and putting natural sympathy aside, to try and get at the rights and wrongs of the business from a higher point of view, namely, that of humanity, which is often very different from that of emotional sentiment.

I ask myself—suppose you knew that by inflicting prolonged pain on 100 rabbits you could discover a way to the extirpation of leprosy, or consumption, or locomotor ataxy, or of suicidal melancholia among human beings, dare you refuse to inflict that pain? Now I am quite unable to say that I dare. That sort of daring would seem to me to be extreme moral cowardice, to involve gross inconsistency.

For the advantage and protection of society, we all agree to inflict pain upon man—pain of the most prolonged and acute character—in our prisons, and on our battlefields. If England were invaded, we should have no hesitation about inflicting the maximum of suffering upon our invaders for no other object than our own good.

But if the good of society and of a nation is a sufficient plea for inflicting pain on men, I think it may suffice us for experimenting on rabbits or dogs.

At the same time, I think that a heavy moral responsibility rests on those who perform experiments of the second kind.

The wanton infliction of pain on man or beast is a crime; pity is that so many of those who (as I think rightly) hold this view, seem to forget that the criminality lies in the wantonness and not in the act of inflicting pain *per se*.—I am, sir, yours faithfully,

T. H. HUXLEY.

So far back as 1870 a committee had been appointed by the British Association, and reported upon the conditions under which they considered experiments on living animals justifiable. In the early spring of 1875 a bill to regulate physiological research was introduced into the Upper House by Lord Hartismere, but not proceeded with. When legislation seemed imminent Huxley, in concert with other men of science, interested himself in drawing up a petition to Parliament to direct opinion on the subject and provide a fair basis for future legislation, which indeed took shape immediately after in a bill introduced by Dr. Lyon Playfair (afterwards Lord Playfair), Messrs. Walpole and Ashley. This bill, though more just to science, did not satisfy many scientific men, and was withdrawn upon the appointment of a Royal Commission.

The following letters to Mr. Darwin bear on this period :—

4 MARLBOROUGH PLACE,
Jan. 22, 1875.

MY DEAR DARWIN—I quite agree with your letter about vivisection as a matter of right and justice in the first place, and secondly as the best method of taking the wind out of the enemy's sails. I will communicate with Burdon Sanderson and see what can be done.

My reliance as against —— and her fanatical following is not in the wisdom and justice of the House of Commons, but in the large number of fox-hunters therein. If physiological experimentation is put down by law, hunting, fishing, and shooting, against which a much better case can be made out, will soon follow —Ever yours very faithfully, T. H. HUXLEY.

SOUTH KENSINGTON,
April 21, 1875.

MY DEAR DARWIN—The day before yesterday I met Playfair at the club, and he told me that he had heard from Miss Elliott that *I* was getting up what she called a "Vivisector's Bill," and that Lord Cardwell was very anxious to talk with some of us about the matter.

So you see that there is no secret about our proceedings. I gave him a general idea of what was doing, and he quite confirmed what Lubbock said about the impossibility of any action being taken in Parliament this session

Playfair said he should like very much to know what we proposed doing, and I should think it would be a good thing to take him into consultation.

On my return I found that Pfluger had sent me his memoir with a note such as he had sent to you.

I read it last night, and I am inclined to think that it is a very important piece of work.

He shows that frogs absolutely deprived of oxygen give off carbonic acid for twenty-five hours, and gives very strong reasons for believing that the evolution of carbonic acid by living matter in general is the result of a process of internal rearrangement of the molecules of the living matter, and not of direct oxidation.

His speculations about the origin of living matter are the best I have seen yet, so far as I understand them. But he plunges into the depths of the higher chemistry in which I am by no means at home. Only this I can see, that the paper is worth careful study.—Ever yours faithfully,   T. H. HUXLEY.

31 ROYAL TERRACE, EDINBURGH,
May 19, 1875.

MY DEAR DARWIN—Playfair has sent a copy of his bill to me, and I am sorry to find that its present wording is such as to render it very unacceptable to all teachers of physiology. In discussing the draft with Litchfield I recollect that I insisted strongly on the necessity of allowing demonstrations to students, but I agreed that it would be sufficient to permit such demonstrations only as could be performed under anæsthetics.

The second clause of the bill, however, by the words "for the purpose of new scientific discovery and for no other purpose," absolutely prohibits any kind of demonstration. It would debar me from showing the circulation in the web of a frog's foot or from exhibiting the pulsations of the heart in a decapitated frog.

And by its secondary effect it would prohibit discovery. Who is to be able to make discoveries unless he knows of his own knowledge what has been already made out? It might as well be ruled that a chemical student should begin with organic analysis.

Surely Burdon Sanderson did not see the draft of the bill as it now stands. The Professors here are up in arms about it, and as the papers have associated my name with the bill I shall have to repudiate it publicly unless something can be done. But what in the world is to be done? I have not written to Playfair yet, and shall wait to hear from you before I do. I have an excellent class here, 340 odd, and like the work. Best regards to Mrs. Darwin.—Ever yours faithfully,

T. H. HUXLEY.

31 ROYAL TERRACE, EDINBURGH,
June 5, 1875.

MY DEAR DARWIN—I see I have forgotten to return Playfair's letter, which I enclose. He sent me a copy of his last letter to you, but it did not reach me till some days after my return from London. In the meanwhile I saw him and Lord Cardwell at the House of Commons on Friday (last week).

Playfair seems rather disgusted at our pronunciamento

against the bill, and he declares that both Sanderson and
Sharpey assented to it   What they were dreaming about
I cannot imagine.   To say that no man shall experiment
except for purpose of original discovery is about as
reasonable as to ordain that no man shall swim unless
he means to go from Dover to Calais.

However the Commission is to be issued, and it is
everything to gain time and let the present madness
subside a little.   I vowed I would never be a member of
another Commission if I could help it, but I suppose I
shall have to serve on this.

I am very busy with my lectures, and am nearly half
through.   I shall not be sorry when they are over, as I
have been grinding away now since last October.—With
kindest regards to Mrs. Darwin, ever yours very faithfully,

T. H. HUXLEY.

He was duly asked to serve on the Commission.
Though his lectures in Edinburgh prevented him from
attending till the end of July no difficulty was made
over this, as the first meetings of the Commission,
which began on June 30, were to be devoted to taking
the less controversial evidence.   In accepting his
nomination he wrote to Mr. Cross (afterwards Lord
Cross), at that time Home Secretary : —

If I can be of any service I shall be very glad to act
on the Commission, sympathising as I do on the one hand
with those who abhor cruelty to animals, and, on
the other, with those who abhor the still greater cruelty
to man which is involved in any attempt to arrest the
progess of physiology and of rational medicine.

The other members of the Commission were Lords
Cardwell and Winmarleigh, Mr. W. E. Forster, Sir

J. B. Karslake, Professor Erichssen, and Mr. R. H. Hutton.

The evidence given before the Commission' bore out the view that English physiologists inflicted no more pain upon animals than could be avoided ; but one witness, not an Englishman, and not having at that time a perfect command of the English language, made statements which appeared to the Commission at least to indicate that the witness was indifferent to animal suffering.   Of this incident Huxley writes to Mr. Darwin at the same time as he forwarded a formal invitation for him to appear as a witness before the Commission :—

4 MARLBOROUGH PLACE,
Oct. 30, 1875.

MY DEAR DARWIN—The inclosed tells its own story. I have done my best to prevent your being bothered, but for various reasons which will occur to you I did not like to appear too obstructive, and I was asked to write to you. The strong feeling of my colleagues (and my own I must say also) is that we ought to have your opinions in our minutes.   At the same time there is a no less strong desire to trouble you as little as possible, and under no circumstances to cause you any risk of injury to health.

What with occupation of time, worry and vexation, this horrid Commission is playing the deuce with me.   I have felt it my duty to act as counsel for Science, and was well satisfied with the way things were going.   But on Thursday when I was absent at the Council of the Royal Society —— was examined, and if what I hear is a correct account of the evidence he gave I may as well throw up my brief.

I am told that he openly professed the most entire

indifference to animal suffering, and said he only gave anæsthetics to keep animals quiet!

I declare to you I did not believe the man lived who was such an unmitigated cynical brute as to profess and act upon such principles, and I would willingly agree to any law which would send him to the treadmill.

The impression his evidence made on Cardwell and Forster is profound, and I am powerless (even if I had the desire which I have not) to combat it. He has done more mischief than all the fanatics put together.

I am utterly disgusted with the whole business.—Ever yours,   T. H. HUXLEY.

Of course keep the little article on Species. It is in some American Encyclopædia published by Appleton. And best thanks for your book. I shall study it some day, and value it as I do every line you have written. Don't mention what I have told you outside the circle of discreet Darwindom.

> 4 MARLBOROUGH PLACE,
> *Nov.* 2, 1875.

MY DEAR DARWIN—Our secretary has telegraphed to you to Down, and written to Queen Anne Street.

But to make sure, I send this note to say that we expect you at 13 Delahay Street [1] at 2 o'clock to-morrow. And that I have looked out the highest chair that was to be got for you.[2]—Ever yours very faithfully,

                                  T. H. HUXLEY.

The Commission reported early in 1876, and a few months after Lord Carnarvon introduced a bill intituled "An Act to amend the law relating to Cruelty to Animals." It was a more drastic measure

---

[1] Where the Commission was sitting.

[2] Mr. Darwin was long in the leg. When he came to our house the biggest hassock was always placed in an arm-chair to give it the requisite height for him.

than was demanded. As a writer in *Nature* (1876, p. 248) puts it : "The evidence on the strength of which legislation was recommended went beyond the facts, the report went beyond the evidence, the recommendations beyond the report, and the bill can hardly be said to have gone beyond the recommendations, but rather to have contradicted them."

As to the working of the law, Huxley referred to it the following year in the address, already cited, on "Elementary Instruction in Physiology" (*Coll. Essays*, iii. 310).

But while I should object to any experimentation which can justly be called painful, and while as a member of a late Royal Commission I did my best to prevent the infliction of needless pain for any purpose, I think it is my duty to take this opportunity of expressing my regret at a condition of the law which permits a boy to troll for pike or set lines with live frog bait for idle amusement, and at the same time lays the teacher of that boy open to the penalty of fine and imprisonment if he uses the same animal for the purpose of exhibiting one of the most beautiful and instructive of physiological spectacles—the circulation in the web of the foot. No one could undertake to affirm that a frog is not inconvenienced by being wrapped up in a wet rag and having his toes tied out, and it cannot be denied that inconvenience is a sort of pain. But you must not inflict the least pain on a vertebrated animal for scientific purposes (though you may do a good deal in that way for gain or for sport) without due licence of the Secretary of State for the Home Department, granted under the authority of the Vivisection Act.

So it comes about that, in this year of grace 1877, two persons may be charged with cruelty to animals. One has impaled a frog, and suffered the creature to writhe about in that condition for hours; the other has pained the animal no more than one of us would be pained by tying strings round his fingers and keeping him in the position of a hydropathic patient.   The first offender says, " I did it because I find fishing very amusing," and the magistrate bids him depart in peace—nay, probably wishes him good sport.   The second pleads, " I wanted to impress a scientific truth with a distinctness attainable in no other way on the minds of my scholars," and the magistrate fines him five pounds.

I cannot but think that this is an anomalous and not wholly creditable state of things.

CHAPTER VII

1875-1876

HUXLEY only delivered one address outside his regular work in 1875, on "Some Results of the 'Challenger' Expedition," given at the Royal Institution on January 29. For all through the summer he was away from London, engaged upon the summer course of lectures on Natural History at Edinburgh. This was due to the fact that Professor (afterwards Sir) Wyville Thomson was still absent on the *Challenger* expedition, and Professor Victor Carus, who had acted as his substitute before, was no longer available. Under these circumstances the Treasury granted Huxley leave of absence from South Kensington. His course began on May 3, and ended on July 23, and he thought it a considerable feat to deal with the whole Animal Kingdom in 54 lectures. No doubt both he and his students worked at high pressure, especially when the latter came scantily prepared for the task, like the late Joseph Thomson, afterwards distinguished as an African traveller, who has left an account of his experience in this class.

Thomson's particular weak point was his Greek, and the terminology of the lectures seems to have been a thorn in his side. This account, which actually tells of the 1876 course, occurs on pp. 36 and 37 of his "Life."

The experience of studying personally under Huxley was a privilege to which he had been looking forward with eager anticipation ; for he had already been fascinated with the charm of Huxley's writings, and had received from them no small amount of mental stimulus. Nor were his expectations disappointed. But he found the work to be unexpectedly hard, and very soon he had the sense of panting to keep pace with the demands of the lecturer. It was not merely that the texture of scientific reasoning in the lectures was so closely knit,—although that was a very palpable fact,—but the character of Huxley's terminology was entirely strange to him. It met him on his weakest side, for it presupposed a knowledge of Greek (being little else than Greek compounds with English terminations) and of Greek he had none.

Huxley's usual lectures, he writes, are something awful to listen to. One half of the class, which numbers about four hundred, have given up in despair from sheer inability to follow him. The strain on the attention of each lecture is so great as to be equal to any ordinary day's work. I feel quite exhausted after them. And then to master his language is something dreadful. But, with all these drawbacks, I would not miss them, even if they were ten times as difficult. They are something glorious, sublime !

Again he writes :—

Huxley is still very difficult to follow, and I have been four times in his lectures completely stuck and utterly helpless. But he has given us eight or nine beautiful lectures on the frog. . . . If you only heard a few of the lectures you would be surprised to find that there were so few missing links in the chain of life, from the amœba to the genus homo.

It was a large class, ultimately reaching 353 and breaking the record of the Edinburgh classes without having recourse to the factitious assistance proposed in the letter of May 16.

His inaugural lecture was delivered under what ought to have been rather trying circumstances. On the way from London he stopped a night with his old friends, John Bruce and his wife (one of the Fannings), at their home, Barmoor Castle, near Beal. He had to leave at 6 next morning, reaching Edinburgh at 10, and lecturing at 2. "Nothing," he writes, "could be much worse, but I am going through it with all the cheerfulness of a Christian martyr."

On May 3 he writes to his wife from the Bruces' Edinburgh house, which they had lent him.

I know that you will be dying to hear how my lecture went off to-day—so I sit down to send you a line, though you did hear from me to-day.

The theatre was crammed. I am told there were 600 auditors, and I could not have wished for more thorough attention. But I had to lecture in gown and Doctor's hood and the heat was awful. The Principal and the chief Professors were present, and altogether it was a state affair. I was in great force, although I did get up at six this morning and travelled all the way from Barmoor. But I won't do that sort of thing again, it's tempting Providence.

May 5.—Fanny and her sisters and the Governess

flit to Barmoor to-day and I shall be alone in my glory
I shall be very comfortable and well cared for, so
make your mind easy, and if I fall ill I am to send
for Clark. He expressly told me to do so as I left
him !

I gave my second lecture yesterday to an audience
filling the theatre. The reason of this is that everybody
who likes—comes for the first week and then only those
who have tickets are admitted. How many will become
regular students I don't know yet, but there is promise
of a big class. The Lord send three extra—to make up
for . . . (a sudden claim upon his purse before he left
home).

And he writes of this custom to Professor Baynes
on June 12 :—

My class is over 350 and I find some good working
material among them. Parsons mustered strong in the
first week, but I fear they came to curse and didn't
remain to pay.

He was still Lord Rector of Aberdeen University,
and on May 10 writes how he attended a business
meeting there :—

I have had my run to Aberdeen and back—got up at
5, started from Edinburgh at 6.25, attended the meeting
of the Court at 1. Then drove out with Webster to
Edgehill in a great storm of rain and was received with
their usual kindness. I did not get back till near 8
o'clock last night and, thanks to *The Virginians* and a
good deal of Virginia, I passed the time pleasantly
enough . . . There are 270 tickets gone up to this date,
so I suppose I may expect a class of 300 men. 300 × 4
= 1200. Hooray.

To his Eldest Daughter

EDINBURGH, *May* 16, 1875.

MY DEAREST JESS—Your mother's letter received this morning reminds me that I have not written to " Cordelia " (I suppose she means Goneril) by a message from that young person—so here is reparation.

I have 330 students, and my class is the biggest in the University—but I am quite cast down and discontented because it is not 351,—being one more than the Botany Class last year—which was never so big before or since.

I am thinking of paying 21 street boys to come and take the extra tickets so that I may crow over all my colleagues.

Fanny Bruce is going to town next week to her grandmother's and I want you girls to make friends with her. It seems to me that she is very nice—but that is only a fallible man's judgment, and Heaven forbid that I should attempt to forestall Miss Cudberry's decision on such a question. Anyhow she has plenty of energy and, among other things, works very hard at *German*.

M——says that the Rootle-Tootles have a bigger drawing-room than ours. I should be sorry to believe these young beginners guilty of so much presumption, and perhaps you will tell them to have it made smaller before I visit them.

A Scotch gentleman has just been telling me that May is the worst month in the year, here; so pleasant ! but the air is soft and warm to-day, and I look out over the foliage to the castle and don't care.

Love to all, and specially M——. Mind you don't tell her that I dine out to-day and to-morrow—positively for the first and last times.—Ever your loving father,

T. H. HUXLEY.

However, the class grew without such adventitious aid, and he writes to Mr. Herbert Spencer on June 15 :—

. . . I have a class of 353, and instruct them in dry facts—particularly warning them to keep free of the infidel speculations which are current under the name of evolution.

I expect an "examiner's call" from a Presbytery before the course is over, but I am afraid that the pay is not enough to induce me to forsake my "larger sphere of influence" in London.

In the same letter he speaks of a flying visit to town which he was about to make on the following Thursday, returning on the Saturday for lack of a good Sunday train :

Mayhap I may chance to see you at the club—but I shall be torn to pieces with things to do during my two days' stay.

If Moses had not existed I should have had three days in town, which is a curious concatenation of circumstances.

As for his health during this period, it maintained, on the whole, a satisfactory level, thanks to the regime of which he writes to Professor Baynes :

I am very sorry to hear that you have been so seriously ill. You will have to take to my way of living—a mutton chop a day and no grog, but much baccy. Don't begin to pick up your threads too fast.

No wonder you are uneasy if you have crabs on your conscience.[1] Thank Heaven they are not on mine !

---

[1] *I.e.* an article for the *Encyclopædia Britannica.*

I am glad to hear you are getting better, and I
sincerely trust that you may find all the good you seek
in the baths.

As to coming back a "new man," who knows what
that might be? Let us rather hope for the old man in
a state of complete repair—A1 copper bottomed.

Excuse my nautical language.

The following letters also touch on his Edinburgh
lectures :—

CRAGSIDE, MORPETH, *August* 11, 1875.

MY DEAR FOSTER—We are staying here with Sir W
Armstrong—the whole brood—Miss Matthaei and the
majority of the chickens being camped at a farm-house
belonging to our host about three miles off. It is wetter
than it need be, otherwise we are very jolly.

I finished off my work in Edinburgh on the 23rd and
positively polished off the Animal Kingdom in 54 lectures.
French without a master in twelve lessons is nothing to
this feat. The men worked very well on the whole, and
sent in some creditable examination papers. I stayed a
few days to finish up the abstracts of my lectures for the
Medical Times; then picked up the two elder girls who
were at Barmoor and brought them on here to join the
wife and the rest.

How is it that Dohrn has been and gone? I have
been meditating a letter to him for an age. He wanted
to see me, and I did not know how to manage to bring
about a meeting.

Edinburgh is greatly exercised in its mind about the
vivisection business, and "Vagus" "swells wisibly" whenever the subject is mentioned. I think there is an
inclination to regard those who are ready to consent to
legislation of any kind as traitors, or, at any rate,
trimmers. It sickens me to reflect on the quantity of
time and worry I shall have to give to that subject when
I get back.

I see that —— has been blowing the trumpet at the Medical Association. He has about as much tact as a flyblown bull.

I have just had a long letter from Wyville Thomson. The *Challenger* inclines to think that *Bathybius* is a mineral precipitate! in which case some enemy will probably say that it is a product of my precipitation. So mind, I was the first to make that "goak." Old Ehrenberg suggested something of the kind to me, but I have not his letter here. I shall eat my leek handsomely, if any eating has to be done. They have found pseudopodia in *Globigerina*.

With all good wishes from ours to yours—Ever yours faithfully,   T. H. HUXLEY.

CRAGSIDE, MORPETH, *August* 13, 1875.

MY DEAR TYNDALL—I find that in the midst of my work in Edinburgh I omitted to write to De Vrij, so I have just sent him a letter expressing my pleasure in being able to co-operate in any plan for doing honour to old Benedict,[1] for whom I have a most especial respect.

I am not sure that I won't write something about him to stir up the Philistines.

My work at Edinburgh got itself done very satisfactorily, and I cleared about £1000 by the transaction, being one of the few examples known of a Southern coming north and pillaging the Scots. However, I was not sorry when it was all over, as I had been hard at work since October and began to get tired.

The wife and babies from the south, and I from the north, met here a fortnight ago and we have been idling very pleasantly ever since. The place is very pretty and our host kindness itself. Miss Matthaei and five of the bairns are at Cartington—a moorland farmhouse three miles off—and in point of rosy cheeks and appetites might com-

---

[1] Spinoza, a memorial to whom was being raised in Holland.

pete with any five children of their age and weight. Jess and Mady are here with us and have been doing great execution at a ball at Newcastle. I really don't know myself when I look at these young women, and my hatred of possible sons-in-law is deadly. All send their love.—Ever yours very faithfully, T. H. HUXLEY.

Wish you joy of Bristol.

The following letter to Darwin was written when the Polar Expedition under Sir George Nares was in preparation. It illustrates the range of observation which his friends had learned to expect in him :—

ATHENÆUM CLUB, *Jan.* 22, 1875.

MY DEAR DARWIN—I write on behalf of the Polar Committee of the Royal Society to ask for any suggestions you may be inclined to offer us as instructions to the naturalists who are to accompany the new expedition.

The task of drawing up detailed instructions is divided among a lot of us; but you are as full of ideas as an egg is full of meat, and are shrewdly suspected of having, somewhere in your capacious cranium, a store of notions which would be of great value to the naturalists.

All I can say is, that if you have not already "collated facts" on this topic, it will be the first subject I ever suggested to you on which you had not.

Of course we do not expect you to put yourself to any great trouble—nor ask for such a thing—but if you will jot down any notes that occur to you we shall be thankful.

We must have everything in hand for printing by March 15.—Ever yours very faithfully,

T. H. HUXLEY.

The following letter dates from soon after the death of Charles Kingsley :—

SCIENCE SCHOOLS, S. KENSINGTON,
Oct. 22, 1875.

DEAR MISS KINGSLEY—I sincerely trust that you believe I have been abroad and prostrated by illness, and have thereby accounted for receiving no reply to your letter of a fortnight back.

The fact is that it has only just reached me, owing to the neglect of the people in Jermyn Street, who ought to have sent it on here.

I assure you I have not forgotten the brief interview to which you refer, and I have often regretted that the hurry and worry of life (which increases with the square of your distance from youth) never allowed me to take advantage of your kind father's invitation to become better acquainted with him and his. I found his card in Jermyn Street when I returned last year, with a pencilled request that I would call on him at Westminster.

I meant to do so, but the whirl of things delayed me until, as I bitterly regret, it was too late.

I am not sure that I have any important letter of your father's but one, written to me some fifteen years ago, on the occasion of the death of a child who was then my only son. It was in reply to a letter of my own written in a humour of savage grief. Most likely he burned the letter, and his reply would be hardly intelligible without it. Moreover, I am not at all sure that I can lay my hands upon your father's letter in a certain chaos of papers which I have never had the courage to face for years. But if you wish I will try.

I am very grieved to hear of Mrs. Kingsley's indisposition. Pray make my kindest remembrances to her, and believe me yours very faithfully,   T. H. HUXLEY.

P. S.—By the way, letters addressed to my private residence,

4 MARLBOROUGH PLACE, N.W.,

are sure not to be delayed. And I have another reason

for giving the address—the hope that when you come to Town you will let my wife and daughters make your acquaintance.

His continued interest in the germ-theory and the question of the origin of life (*Address at the British Association*, 1870, see ii. p. 14, *sq.*), appears from the following :—

4 MARLBOROUGH PLACE,
Oct 15, 1875.

MY DEAR TYNDALL—Will you bring with you to the *x* to-morrow a little bottle full of fluid containing the bacteria you have found developed in your infusions? I mean a good characteristic specimen. It will be useful to you, I think, if I determine the forms with my own microscope, and make drawings of them which you can use.—Ever yours,   T. H. HUXLEY.

I can't tell you how delighted I was with the experiments.

Throughout this period, and for some time later, he was in frequent communication with Thomas Spencer Baynes, Professor of Logic and English Literature at St. Andrews University, the editor of the new *Encyclopædia Britannica*, work upon which was begun at the end of 1873. From the first Huxley was an active helper, both in classifying the biological subjects which ought to be treated of, suggesting the right men to undertake the work, and himself writing several articles, notably that on Evolution.[1]

---

[1] Others were *Actinozoa*, *Amphibia*, *Animal Kingdom*, and *Biology*.

Extracts from his letters to Professor Baynes between the years 1873 and 1884, serve to illustrate the work which he did and the relations he maintained with the genial and learned editor.

Nov. 2, 1873.—I have been spending my Sunday morning in drawing up a list of headings, which will I think exhaust biology from the Animal point of view, and each of which does not involve more than you are likely to get from one man.   In many cases, *i.e. Insecta, Entomology,* I have subdivided the subjects, because, by an unlucky peculiarity of workers in these subjects, men who understand zoology from its systematic side are often ignorant of anatomy, and those who know fossils are often weak in recent forms.

But of course the subdivision does not imply that one man should not take the whole if he is competent to do so.   And if separate contributors supply articles on these several subdivisions, somebody must see that they work in harmony.

But with all the good will in the world, he was too hard pressed to get his quota done as quickly as he wished.   He suggests at once that "Hydrozoa" and "Actinozoa," in his list, should be dealt with by the writer of the article "Cœlenterata."

Shunting "Actinozoa" to "Cœlenterata" would do no harm, and would have the great merit of letting me breathe a little.   But if you think better that "Actinozoa" should come in its place under A, I will try what I can do.

December 30, 1873—As to *Anthropology,* I really am afraid to promise.   At present I am plunged in *Amphibia,* doing a lot of original work to settle questions which

have been hanging vaguely in my mind for years. If
Amphibia is done by the end of January it is as much as
it will be.

In February I must give myself—or at any rate my
spare self—up to my Rectorial Address,[1] which (tell it
not in Gath) I wish at the bottom of the Red Sea. And
I do not suppose I shall be able to look seriously at either
Animal Kingdom or *Anthropology* before the address is
done with. And all depends on the centre of my micro-
cosm—intestinum colon—which plays me a trick every
now and then.

I will do what I can if you like, but if you trust me
it is at your proper peril.

Feb. 8, 1874.—How astonished folks will be if eloquent
passages out of the address get among the *Amphibia*, and
comments on Frog anatomy into the address. As I am
working at both just now this result is not improbable.

Meanwhile the address and the ten days' stay at
Aberdeen had been "playing havoc with the
Amphibia," but on returning home, he went to work
upon the latter, and writes on March 12 :—

I did not care to answer your last letter until I had
an instalment of *Amphibia* ready. Said instalment was
sent off to you, care of Messrs. Black, yesterday, and now
I feel like Dick Swiveller, when happy circumstances
having enabled him to pay off an old score he was able
to begin running up another.

June 8.—I have had sundry proofs and returned
them. My writing is lamentable when I am in a hurry,
but I never provoked a strike before ! I declare I think
I write as well as the editor, on ordinary occasions.

He was pleased to find someone who wrote as
badly as, or worse than, himself, and several times

---

[1] His Rectorial Address at Aberdeen. (See above, p. 122.)

rallies Baynes on that score. Thus, when Mrs. Baynes had acted as her husband's amanuensis, he writes (February 11, 1878) :—

My respectful compliments to the "mere machine," whose beautiful caligraphy (if that isn't a tautology) leaves no doubt in my mind that whether the writing of your letters by that agency is good for you or not it is admirable for your correspondents.

Why people can't write a plain legible hand I can't imagine.[1]

And on another occasion he adds a postscript to say, "You write worse than ever. So do I."

However, the article got finished in course of time :—

Aug. 5.—I have seen and done with all *Amphibia* but the last sheet, and that only waits revise. Considering it was to be done in May, I think I am pretty punctual.

The next year, immediately before taking Sir Wyville Thomson's lectures at Edinburgh, he writes about another article which he had in hand :—

4 MARLBOROUGH PLACE, N.W.,
March 16, 1875.

MY DEAR BAYNES—I am working against time to get a lot of things done—amongst others BIOLOGY—before I go north. I have written a large part of said article, and it would facilitate my operation immensely if what is done were set up and I had two or three proofs, one for Dyer, who is to do part of the article.

Now, if I send the MS. to North Bridge will you

---

[1] *N.B.*—This sentence is written purposely in a most illegible hand.

swear by your gods (0—1—3—1 or any greater number
as the case may be) that I shall have a proof swiftly and
not be kept waiting for weeks till the whole thing has
got cold, and I am at something else a hundred miles
away from Biology?

If not I will keep the MS. till it is all done, and you
know what that means.—Ever yours very truly,

T. H. HUXLEY.

CRAGSIDE, MORPETH,
Aug. 12, 1875.

MY DEAR BAYNES—The remainder of the proof of
"Biology" is posted to-day—"Praise de Lor'."

I have a dim recollection of having been led by your
soft and insinuating ways to say that I would think
(only *think*) about some other article. What the deuce
was it?

I have told the Royal Society people to send you a
list of Fellows, addressed to Black's.

We have had here what may be called bad weather
for England, but it has been far better than the best
Edinburgh weather known to my experience.

All my friends are out committing grouse-murder.
As a vivisection Commissioner I did not think I could
properly accompany them.—Ever yours very faithfully,

T. H. HUXLEY.

CRAGSIDE, MORPETH,
Aug. 24, 1875.

MY DEAR BAYNES—I think —— is like enough to
do the "Cœlenterata" well if you can make sure of his
doing it at all. He is a man of really great knowledge
of the literature of Zoology, and if it had not been for
the accident of being a procrastinating impracticable ass,
he could have been a distinguished man. But he is a
sort of Balaam-Centaur with the asinine stronger than
the prophetic moiety.

I should be disposed to try him, nevertheless.

I don't think I have had final revise of Biology yet.

I do not know that "Cœlenterata" is Lankester's speciality. However, he is sure to do it well if he takes it up.—Ever yours very faithfully,   T. H. HUXLEY.

4 MARLBOROUGH PLACE, N.W.,
Oct. 12, 1875.

MY DEAR BAYNES—Do you remember my telling you that I should before long be publishing a book, of which general considerations on Biology would form a part, and that I should have to go over the same ground as in the article for the Encyclopædia?

Well, that prediction is about to be verified, and I want to know what I am to do.

You see, as I am neither dealing with Theology, nor History, nor Criticism, I can't take a fresh departure and say something entirely different from what I have just written.

On the other hand, if I republish what stands in the article, the Encyclopædia very naturally growls.

What do the sweetest of Editors and the most liberal of Proprietors say ought to be done under the circumstances?

I pause for a reply.

I have carried about Stanley's [1] note in my pocketbook until I am sorry to say the flyleaf has become hideously stained.

The wife and daughters could make nothing of it, but I, accustomed to the MS. of certain correspondents, have no doubt as to the fourth word of the second sentence. It is "Canterbury." [2] Nothing can be plainer.

Hoping the solution is entirely satisfactory,—Believe me, ever yours very faithfully,   T. H. HUXLEY.

---

[1] The Dean's handwriting was proverbial.

[2] The writing of this word is carefully slurred until it is almost as illegible as the original.

Though he refused to undertake the article on *Distribution*, he managed to write that on *Evolution* (republished in *Collected Essays*, ii. 187). Thus on July 28, 1877, he writes :—

I ought to do "Evolution," but I mightn't and I shouldn't. Don't see how it is practicable to do justice to it with the time at my disposal, though I really should like to do it, and I am at my wits' end to think of anybody who can be trusted with it.

Perhaps something may turn up, and if so I will let you know.

The something in the way of more time did turn up by dint of extra pressure, and the article got written in the course of the autumn, as appears from the following of December 29, 1877 :—

I send you the promised skeleton (with a good deal of the flesh) of Evolution. It is costing me infinite labour in the way of reading, but I am glad to be obliged to do the work, which will be a curious and instructive chapter in the history of Science.

The lawyer-like faculty of putting aside a subject when done with, which is indicated in the letter of March 16, 1875, reappears in the following :—

4 MARLBOROUGH PLACE, N.W.,
March 18, 1878.

MY DEAR BAYNES—Your printers are the worst species of that diabolic genus I know of. It is at least a month since I sent them a revise of "Evolution" by no means finished, and from that time to this I have had nothing from them.

I shall forget all about the subject, and then at the

last moment they will send me a revise in a great hurry, and expect it back by return of post.

But if they get it, may I go to their Father !—Ever yours very faithfully,   T. H. HUXLEY.

Later on, the pressure of work again forbade him to undertake further articles on *Harvey, Hunter,* and *Instinct.*

I am sorry to say that my hands are full, and I have sworn by as many gods as Hume has left me, to undertake nothing more for a long while beyond what I am already pledged to do, a small book anent Harvey being one of these things.

And on June 9 :—

After nine days' meditation (directed exclusively to the Harvey and Hunter question) I am not any "forrarder," as the farmer said after his third bottle of Gladstone claret. So perhaps I had better mention the fact. I am very glad you have limed Flower for " Mammalia " and " Horse "—nobody could be better.

4 MARLBOROUGH PLACE, N.W.,
July 1, 1879.

MY DEAR BAYNES—On Thursday last I sought for you at the Athenæum in the middle of the day, and told them to let me know if you came in the evening when I was there again. But I doubt not you were plunged in dissipation.

My demonstrator Parker showed me to-day a letter he had received from Black's, asking him to do anything in the small Zoology way between H and L.

He is a modest man, and so didn't ask what the H——L he was to do, but he looked it.

Will you enlighten him or me, and I will convey the information on ?

I had another daughter married yesterday. She was a great pet and it is very hard lines on father and mother. The only consolation is that she has married a right good fellow, John Collier the artist.—Ever yours very faithfully, T. H. HUXLEY.

July 19, 1879.—Many thanks for your and Mrs. Baynes' congratulations. I am very well content with my son-in-law, and have almost forgiven him for carrying off one of my pets, which shows a Christian spirit hardly to be expected of me.

SOUTH KENSINGTON,
July 2, 1880.

MY DEAR BAYNES—I have been thinking over the matter of Instinct, and have come to the conclusion that I dare not undertake anything fresh.

There is an address at Birmingham in the autumn looming large, and ghosts of unfinished work flitter threateningly.—Ever yours very faithfully,

T. H. HUXLEY.

CHAPTER VIII

1876

THE year 1876 was again a busy one, almost as busy as any that went before. As in 1875, his London work was cut in two by a course of lectures in Edinburgh, and sittings of the Royal Commission on Scottish Universities, and furthermore, by a trip to America in his summer vacation.

In the winter and early spring he gave his usual lectures at South Kensington; a course to working men "On the Evidence as to the Origin of Existing Vertebrated Animals," from February to April (*Nature*, vols. xiii. and xiv.) ; a lecture at the Royal Institution (January 28) "On the Border Territory between the Animal and Vegetable Kingdoms" (*Coll. Essays*, viii. 170); and another at Glasgow (February 15) "On the Teleology and Morphology of the Hand."

In this lecture, which he never found time to get into final shape for publication, but which was substantially repeated at the Working Men's College in 1878, he touched upon one of the philosophic aspects of the theory of evolution, namely, how far is it consistent with the argument from design?

Granting provisionally the force of Paley's argument in individual cases of adaptation, and illustrating it by the hand and its representative in various of the Mammalia, he proceeds to show by the facts of morphology that the argument, as commonly stated, fails; that each mechanism, each animal, was not specially made to suit the particular purpose we find it serving, but was developed from a single common type. Yet in a limited and special sense he finds teleology to be not inconsistent with morphology. The two sets of facts flow from a common cause, evolution. Descent by modification accounts for similarity of structure; the process of gradual adaptation to conditions accounts for the existing adaptation to purpose. To be a teleologist and yet accept evolution it is only necessary "to suppose that the original plan was sketched out—that the purpose was foreshadowed in the molecular arrangements out of which the animals have come."

This was no new view of his. While, ever since his first review of the *Origin* in 1859 (*Coll. Ess.* ii. 6), he had declared the commoner and coarser forms of teleology to find their most formidable opponent in the theory of evolution, and in 1869, addressing the Geological Society, had spoken of "those final causes, which have been named barren virgins, but which might be more fitly termed the *hetairœ* of philosophy, so constantly have they led men astray" (*ib.* viii. 80;

cp. ii. 21, 36), he had, in his *Criticism of the Origin* (1864, ii. 86), and the *Genealogy of Animals* (1869, ii. 109, *sqq.*), shown how "perhaps the most remarkable service to the philosophy of Biology rendered by Mr. Darwin is the reconciliation of teleology and morphology, and the explanation of the facts of both which his views offer . . . the wider teleology, which is actually based upon the fundamental proposition of evolution."

His note-book shows that he was busy with Reptilia from Elgin and from India; and with his *Manual of Invertebrate Anatomy*, which was published the next year; while he refused to undertake a course of ten lectures at the Royal Institution, saying that he had already too much other work to do, and would have no time for original work.

About this time, also, in answer to a request from a believer in miracles, "that those who fail to perceive the cogency of the evidence by which the occurrence of miracles is supported, should not confine themselves to the discussion of general principles, but should grapple with some particular case of an alleged miracle," he read before the Metaphysical Society a paper dealing with the evidence for the miracle of the resurrection. (See i. p. 459.)

Some friends wished him to publish the paper as a contribution to criticism; but his own doubts as to the opportuneness of so doing were confirmed by a letter from Mr. John Morley, then editor of the *Fortnightly Review*, to which he replied (January 18):—

To say truth, most of the considerations you put so forcibly had passed through my mind—but one always suspects oneself of cowardice when one's own interests may be affected.

At the beginning of May he went to Edinburgh. He writes home on May 8 :—

I am in hopes of being left to myself this time, as nobody has called but Sir Alexander Grant the Principal, Crum Brown, whom I met in the street just now, and Lister, who has a patient in the house. I have been getting through an enormous quantity of reading, some tough monographs that I brought with me, the first volume of Forster's *Life of Swift*, *Goodsir's Life*, and a couple of novels of George Sand, with a trifle of Paul Heyse. You should read George Sand's *Césarine Dietrich* and *La Mare au Diable* that I have just finished. She is bigger than George Eliot, more flexible, a more thorough artist. It is a queer thing, by the way, that I have never read *Consuelo*. I shall get it here. When I come back from my lecture I like to rest for an hour or two over a good story. It freshens me wonderfully.

However, social Edinburgh did not leave him long to himself, but though he might thus lose something of working time, this loss was counterbalanced by the dispelling of some of the fits of depression which still assailed him from time to time.

On May 25 he writes :—

The General Assembly is sitting now, and I thought I would look in. It was very crowded and I had to stand, so I was soon spied out and invited to sit beside the Lord High Commissioner, who represents the Crown in the Assembly, and there I heard an ecclesiastical row about whether a certain church should be allowed to

have a cover with IHS on the Communion Table or not.
After three hours' discussion the IHSers were beaten. I
was introduced to the Commissioner Lord Galloway, and
asked to dine to-night. So I felt bound to go to the
special levee at Holyrood with my colleagues this morning,
and I shall have to go to my Lady Galloway's reception
in honour of the Queen's birthday to-morrow. Luckily
there will be no more of it. Vanity of Vanities!
Saturday afternoon I go out to Lord Young's place to
spend Sunday. I have been in rather a hypochondriacal
state of mind, and I will see if this course of medicine
will drive the seven devils out.

One of the chief friendships which sprang from
this residence in Edinburgh was that with Dr. (afterwards Sir John) Skelton, widely known under his
literary pseudonym of "Shirley." A Civil Servant
as well as a man of letters, he united practical life
with literature, a combination that appealed particularly to Huxley, so that he was a constant visitor
at Dr. Skelton's picturesque house, the Hermitage
of Braid, near Edinburgh. A number of letters
addressed to Skelton from 1875 to 1891 show that
with him Huxley felt the stimulus of an appreciative
correspondent.

4 MELVILLE STREET, EDINBURGH,
June 23, 1876.

MY DEAR SKELTON—I do not understand how it is
that your note has been so long in reaching me ; but I
hasten to repel the libellous insinuation that I have
vowed a vow against dining at the Hermitage.

I wish I could support that repudiation by at once
accepting your invitation for Saturday or Sunday, but

my Saturdays and Sundays are mortgaged to one or other of your judges (good judges, obviously).

Shall you be at home on Monday or Tuesday? If so, I would put on a kilt (to be as little dressed as possible), and find my way out and back; happily improving my mind on the journey with the tracts you mention.—Ever yours very faithfully,

T. H. HUXLEY.

4 MELVILLE STREET, EDINBURGH,
July 1, 1876.

MY DEAR SKELTON—Very many thanks for the copy of the *Comedy of the Noctes*, which reached me two or three days ago. Turning over the pages I came upon the Shepherd's "Terrible Journey of Timbuctoo," which I enjoyed as much as when I first read it thirty odd years ago.—Ever yours very faithfully, T. H. HUXLEY.

On June 23 he writes home :—

Did you read Gilman's note asking me to give the inaugural discourse at the Johns Hopkins University, and offering £100 on the part of the trustees? I am minded to do it on our way back from the south, but don't much like taking money for the performance. Tell me what you think about this at once, as I must reply.

This visit to America had been under discussion for some time. It is mentioned as a possibility in a letter to Darwin two years before. Early in 1876 Mr. Frederic Harrison was commissioned by an American correspondent—who, by the way, had named his son Thomas Huxley—to give my father the following message :—"The whole nation is electrified by the announcement that Professor Huxley is to visit us next fall. We will make infinitely more of

him than we did of the Prince of Wales and his
retinue of lords and dukes." Certainly the people
of the States gave him an enthusiastic welcome; his
writings had made him known far and wide; as the
manager of the Californian department at the Philadelphia Exhibition told him, the very miners of
California read his books over their camp fires; and
his visit was so far like a royal progress, that unless
he entered a city disguised under the name of Jones
or Smith, he was liable not merely to be interviewed,
but to be called upon to "address a few words" to
the citizens.

Leaving their family under the hospitable care of
Sir W. and Lady Armstrong at Cragside, my father
and mother started on July 27 on board the *Germanic*,
reaching New York on August 5. My father sometimes would refer, half-jestingly, to the trip as his
second honeymoon, when, for the first time in twenty
years, he and my mother set forth by themselves,
free from all family cares. And indeed, there was
the underlying resemblance that this too came at the
end of a period of struggle to attain, and marked the
beginning of a more settled period. His reception
in America may be said to emphasise his definite
establishment in the first rank of English thinkers.
It was a signal testimony to the wide extent of his
influence, hardly suspected, indeed, by himself; an
influence due above all to the fact that he did not
allow his studies to stand apart from the moving
problems of existence, but brought the new and

regenerating ideas into contact with life at every point, and that his championship of the new doctrines had at the same time been a championship of freedom and sincerity in thought and word against shams and self-deceptions of every kind. It was not so much the preacher of new doctrines who was welcomed, as the apostle of veracity—not so much the student of science as the teacher of men.

Moreover, another sentiment coloured this holiday visit. He was to see again the beloved sister of his boyhood. She had always prophesied his success, and now after thirty years her prophecy was fulfilled by his coming, and, indeed, exceeded by the manner of it.

Mr. Smalley, then London correspondent of the *New York Tribune*, was a fellow passenger of his on board the *Germanic*, and tells an interesting anecdote of him :—

Mr. Huxley stood on the deck of the *Germanic* as she steamed up the harbour of New York, and he enjoyed to the full that marvellous panorama. At all times he was on intimate terms with Nature and also with the joint work of Nature and Man ; Man's place in Nature being to him interesting from more points of view than one. As we drew near the city—this was in 1876, you will remember—he asked what were the tall tower and tall building with a cupola, then the two most conspicuous objects. I told him the Tribune and the Western Union Telegraph buildings. "Ah," he said, "that is interesting; that is American. In the Old World the first things you see as you approach a great city are steeples ; here you see, first, centres of intelligence." Next to those the tug-boats seemed to attract him as they tore fiercely up and down and across the bay. He looked long at them and finally said, "If I were not a man I think I should like to be a tug." They seemed to him the condensation and complete expression of the energy and force in which he delighted.

The personal welcome he received from the friends he visited was of the warmest. On the arrival of the *Germanic* the travellers were met by Mr. Appleton the publisher, and carried off to his country house at Riverdale. While his wife was taken to Saratoga to see what an American summer resort was like, he himself went on the 9th to New Haven, to inspect the fossils at Yale College, collected from the Tertiary deposits of the Far West by Professor Marsh, with great labour and sometimes at the risk of his scalp. Professor Marsh told me how he took him to the University, and proposed to begin by showing him over the buildings. He refused. "Show me what you have got inside them; I can see plenty of bricks and mortar in my own country." So they went straight to the fossils, and as Professor Marsh writes :—[1]

One of Huxley's lectures in New York was to be on the genealogy of the horse, a subject which he had already written about, based entirely upon European specimens. My own explorations had led me to conclusions quite different from his, and my specimens seemed to me to prove conclusively that the horse originated in the New World and not in the Old, and that its genealogy must

---

[1] *American Journal of Science*, vol. l. August 1895.

be worked out here. With some hesitation, I laid the whole matter frankly before Huxley, and he spent nearly two days going over my specimens with me, and testing each point I made.

At each enquiry, whether he had a specimen to illustrate such and such a point or exemplify a transition from earlier and less specialised forms to later and more specialised ones, Professor Marsh would simply turn to his assistant and bid him fetch box number so and so, until Huxley turned upon him and said, "I believe you are a magician ; whatever I want, you just conjure it up."

The upshot of this examination was that he recast a great part of what he meant to say at New York. When he had seen the specimens, and thoroughly weighed their import, continues Professor Marsh—

He then informed me that all this was new to him, and that my facts demonstrated the evolution of the horse beyond question, and for the first time indicated the direct line of descent of an existing animal. With the generosity of true greatness, he gave up his own opinions in the face of new truth, and took my conclusions as the basis of his famous New York lecture on the horse. He urged me to prepare without delay a volume on the genealogy of the horse, based upon the specimens I had shown him. This I promised, but other work and new duties have thus far prevented.

A letter to his wife describes his visit to Yale :—

My excellent host met me at the station, and seems as if he could not make enough of me. I am installed in apartments which were occupied by his uncle, the millionaire Peabody, and am as quiet as if I were in my own house. We have had a preliminary canter over the fossils, and I have seen some things which were worth all the journey across.

This is the most charmingly picturesque town, with the streets lined by avenues of elm trees which meet overhead. I have never seen anything like it, and you must come and look at it. There is fossil work enough to occupy me till the end of the week, and I have arranged to go to Springfield on Monday to examine the famous footprints of the Connecticut Valley.

The Governor has called upon me, and I shall have to go and do pretty-behaved *chez lui* to-morrow. An application has come for an autograph, but I have not been interviewed !

This immunity, however, did not last long. He appears to have been caught by the interviewer the next day, for he writes on the 11th :—

I have not seen the notice in the *World* you speak of. You will be amused at the article written by the interviewer. He was evidently surprised to meet with so little of the "highfalutin" philosopher in me, and says I am "affable" and of "the commercial or mercantile" type. That is something I did not know, and I am rather proud of it. We may be rich yet.

As to his work at Yale Museum, he writes in the same letter :—

We are hard at work still. Breakfast at 8.30—go over to the Museum with Marsh at 9 or 10—work till 1.30—dine—go back to Museum to work till 6. Then Marsh takes me for a drive to see the views about the town, and back to tea about half-past eight. He is a wonderfully good fellow, full of fun and stories about his Western adventures, and the collection of fossils is the most wonderful thing I ever saw. I wish I could spare three weeks instead of one to study it.

To-morrow evening we are to have a dinner by way of winding up, and he has asked a lot of notables to meet me. I assure you I am being "made of," as I thought nobody but the little wife was foolish enough to do.

On the 16th he left to join Professor Alexander Agassiz at Newport, whence he wrote the following letters :—

NEWPORT, *Aug.* 17, 1876.

MY DEAR MARSH—I really cannot say how much I enjoyed my visit to New Haven. My recollections are sorting themselves out by degrees and I find how rich my store is. The more I think of it the more clear it is that your great work is the settlement of the pedigree of the horse.

My wife joins with me in kind regards. I am yours very faithfully,   T. H. HUXLEY.

TO MR. CLARENCE KING

NEWPORT, *Aug.* 19, 1876.

MY DEAR SIR—In accordance with your wish, I very willingly put into writing the substance of the opinion as to the importance of Professor Marsh's collection of fossils which I expressed to you yesterday. As you are aware, I devoted four or five days to the examination of this collection, and was enabled by Prof. Marsh's kindness to obtain a fair conception of the whole.

I am disposed to think that whether we regard the abundance of material, the number of complete skeletons of the various species, or the extent of geological time covered by the collection, which I had the good fortune to see at New Haven, there is no collection of fossil vertebrates in existence which can be compared with it. I say this without forgetting Montmartre, Siwalik, or Pikermi—and I think that I am quite safe in adding that no collection which has been hitherto formed approaches that made by Professor Marsh, in the completeness of the chain of evidence by which certain existing mammals are connected with their older tertiary ancestry.

It is of the highest importance to the progress of Biological Science that the publication of this evidence, accompanied by illustrations of such fulness as to enable palæontologists to form their own judgment as to its value, should take place without delay.—I am yours very faithfully,   THOMAS H. HUXLEY.

Breaking their journey at Boston, they went from Newport to Petersham, in the highlands of Worcester County, where they were the guests of Mr. and Mrs. John Fiske, at their summer home.   Among the other visitors were the eminent musical composer Mr. Paine, the poet Cranch, and daughters of Hawthorne and Longfellow, so that they found themselves in the midst of a particularly cheerful and delightful party. From Petersham they proceeded to Buffalo, the meeting-place that year of the American Association for the Advancement of Science, which my father had promised to attend.   Here they stayed with Mr. Marshall, a leading lawyer, who afterwards visited them in England.

A week was spent at Niagara, partly in making holiday, partly in shaping the lectures which had to be delivered at the end of the trip.   As to the impression made upon him by the Falls—an experience which, it is generally presumed, every traveller is bound to record—I may note that after the first disappointment at their appearance, inevitable wherever the height of a waterfall is less than the breadth, he found in them an inexhaustible charm and fascination. As in duty bound, he, with my mother, completed his experiences by going under the wall of waters to the "Cave of the Winds." But of all things nothing pleased him more than to sit of an evening by the edge of the river, and through the roar of the cataract to listen for the under-sound of the beaten stones grinding together at its foot.

Leaving Niagara on September 2, they travelled to Cincinnati, a 20-hours' journey, where they rested a day; on the 4th another 10 hours took them to Nashville, where they were to meet his sister, Mrs. Scott. Though 11 years his senior, she maintained her vigour and brightness undimmed, as indeed she did to the end of her life, surviving him by a few weeks. As she now stood on the platform at Nashville, Mrs. Huxley, who had never seen her, picked her out from among all the people by her piercing black eyes, so like those of her mother as described in the Autobiographical sketch (*Coll. Ess.* i.).

Nashville, her son's home, had been chosen as the meeting-place by Mrs. Scott, because it was not so far south nor so hot as Montgomery, where she was then living. Nevertheless in Tennessee the heat of the American summer was very trying, and the good people of the town further drew upon the too limited opportunities of their guest's brief visit by sending a formal deputation to beg that he would either deliver an address, or be entertained at a public dinner, or "state his views"—to an interviewer I suppose. He could not well refuse one of the alternatives; and the greater part of one day was spent in preparing a short address on the geology of Tennessee, which was delivered on the evening of September 7. He spoke for twenty minutes, but had scarcely any voice, which was not to be wondered at, as he was so tired that he had kept his room the whole day, while his wife received the endless string of callers.

The next day they returned to Cincinnati; and on the 9th went on to Baltimore, where they stayed with Mr. Garrett, then President of the Baltimore and Ohio railway.

The Johns Hopkins University at Baltimore, for which he was to deliver the opening address, had been instituted by its founder on a novel basis. It was devoted to post-graduate study; the professors and lecturers received incomes entirely independent of the pupils they taught. Men came to study for the sake of learning, not for the sake of passing some future examination. The endowment was devoted in the first place to the furtherance of research; the erection of buildings was put into the background. "It has been my fate," commented Huxley, "to see great educational funds fossilise into mere bricks and mortar in the petrifying springs of architecture, with nothing left to work them. A great warrior is said to have made a desert and called it peace. Trustees have sometimes made a palace and called it a university."

Half the fortune of the founder had gone to this university; the other half to the foundation of a great and splendidly equipped hospital for Baltimore. This was the reason why the discussion of medical training occupies fully half of the address upon the general principles of education, in which, indeed, lies the heart of his message to America, a message already delivered to the old country, but specially appropriate for the new nation developing so rapidly in size and physical resources.

I cannot say that I am in the slightest degree impressed by your bigness or your material resources, as such. Size is not grandeur, territory does not make a nation. The great issue, about which hangs a true sublimity, and the terror of overhanging fate, is, what are you going to do with all these things ? . . .

The one condition of success, your sole safeguard, is the moral worth and intellectual clearness of the individual citizen. Education cannot give these, but it can cherish them and bring them to the front in whatever station of society they are to be found, and the universities ought to be and may be, the fortresses of the higher life of the nation.

This address was delivered under circumstances of peculiar difficulty. The day before, an expedition had been made to Washington, from which Huxley returned very tired, only to be told that he was to attend a formal dinner and reception the same evening. "I don't know how I shall stand it," he

remarked. Going to his room, he snatched an hour
or two of rest, but was then called upon to finish his
address before going out. It seems that it had to be
ready for simultaneous publication in the New York
papers. Now the lecture was not written out; it
was to be given from notes only. So he had to
deliver it *in extenso* to the reporter, who took it down
in shorthand, promising to let him have a longhand
copy in good time the next morning. It did not
come till the last moment. Glancing at it on his
way to the lecture theatre, he discovered to his
horror that it was written upon "flimsy," from which
he would not be able to read it with any success.
He wisely gave up the attempt, and made up his
mind to deliver the lecture as best he could from
memory. The lecture as delivered was very nearly
the same as that which he had dictated the night
before, but with some curious discrepancies between
the two accounts, which, he used to say, occurring as
they did in versions both purporting to have been
taken down from his lips, might well lead the ingenious critic of the future to pronounce them both
spurious, and to declare that the pretended original
was never delivered under the circumstances alleged.[1]

There was an audience of some 2000, and I am
told that when he began to speak of the time that
would come when they too would experience the
dangers of over-population and poverty in their
midst, and would then understand what Europe had

[1] Cp. the incident at Belfast, p. 134.

to contend with more fully than they did, a pin could have been heard to drop. At the end of the lecture, amid the enthusiastic applause of the crowd, he made his way to the front of the box where his hosts and their party were, and received their warm congratulations. But he missed one voice amongst them, and turning to where his wife sat in silent triumph almost beyond speech, he said, "And have you no word for me?" then, himself also deeply moved, stooped down and kissed her.

This address was delivered on Tuesday, September 12. On the 14th he went to Philadelphia, and on the 15th to New York, where he delivered his three lectures on Evolution on Monday, Wednesday, and Friday, September 18, 20, and 22.

These lectures are very good examples of the skill with which he could present a complicated subject in a simple form, the subject seeming to unroll itself by the force of its own naked logic, and carrying conviction the further through the simplicity of its presentation. Indeed, an unfriendly critic once paid him an unintended compliment, when trying to make out that he was no great speaker; that all he did was to set some interesting theory unadorned before his audience, when such success as he attained was due to the compelling nature of the subject itself.

Since his earlier lectures to the public on evolution, the paleontological evidences had been accumulating; the case could be stated without some of the reservations of former days; and he brings forward two telling instances in considerable detail, the one show-ing how the gulf between two such apparently distinct groups as Birds and Reptiles is bridged over by ancient fossils intermediate in form ; the other illus-trating from Professor Marsh's new collections the lineal descent of the specialised Horse from the more general type of quadruped.

The farthest back of these was a creature with four toes on the front limb and three on the hind limb. Judging from the completeness of the series or forms so far, he ventured to indulge in a prophecy.

Thus, thanks to these important researches, it has become evident that, so far as our present knowledge extends, the history of the horse-type is exactly and pre-cisely that which could have been predicted from a knowledge of the principles of evolution. And the knowledge we now possess justifies us completely in the anticipation that when the still lower Eocene deposits, and those which belong to the Cretaceous epoch, have yielded up their remains of ancestral equine animals, we shall find, first, a form with four complete toes and a rudiment of the innermost or first digit in front, with, probably, a rudiment of the fifth digit in the hind foot; while, in still older forms, the series of the digits will be more and more complete, until we come to the five-toed animals, in which, if the doctrine of evolution is well founded, the whole series must have taken its origin.

Seldom has prophecy been sooner fulfilled. Within two months, Professor Marsh had discovered a new genus of equine mammals, Eohippus, from the lowest Eocene deposits of the West, which corresponds very nearly to the description given above.

He continues :—

That is what I mean by demonstrative evidence of evolution. An inductive hypothesis is said to be demonstrated when the facts are shown to be in entire accordance with it. If that is not scientific proof, there are no merely inductive conclusions which can be said to be proved. And the doctrine of evolution, at the present time, rests upon exactly as secure a foundation as the Copernican theory of the motions of the heavenly bodies did at the time of its promulgation. Its logical basis is of precisely the same character—the coincidence of the observed facts with theoretical requirements.

He left New York on September 23. " I had a very pleasant trip in Yankee-land," he writes to Professor Baynes, "and did *not* give utterance to a good deal that I am reported to have said there." He reached England in good time for the beginning of his autumn lectures, and his ordinary busy life absorbed him again. He did not fail to give his London audiences the results of the recent discoveries in American paleontology, and on December 4, delivered a lecture at the London Institution, " On Recent Additions to the Knowledge of the Pedigree of the Horse." In connection with this he writes to Professor Marsh :—

4 MARLBOROUGH PLACE, LONDON, N.W.
Dec. 27, 1876.

MY DEAR MARSH—I hope you do not think it remiss of me that I have not written to you since my return, but you will understand that I plunged into a coil of work, and will forgive me. But I do not mean to let

the year slip away without sending you all our good
wishes for its successor—which I hope will not vanish
without seeing you among us.

I blew your trumpet the other day at the London
Institution in a lecture about the Horse question.   I did
not know then that you had got another step back as I
see you have by the note to my last lecture, which
Youmans has just sent me.

I must thank you very heartily for the pains you
have taken over the woodcuts of the lectures.   It is a
great improvement to have the patterns of the grinders.

I have promised to give a lecture at the Royal Institution on the 21st January next, and I am thinking of
discoursing on the Birds with teeth.   Have you anything
new to tell on that subject ?   I have implicit faith in
the inexhaustibility of the contents of those boxes.

Our voyage home was not so successful as that out.
The weather was cold and I got a chill which laid me up
for several days, in fact I was not well for some weeks
after my return.   But I am vigorous again now.

Pray remember me kindly to all New Haven friends.
My wife joins with me in kindest regards and good
wishes for the new year.   "Tell him we expect to see
him next year."—I am, yours very faithfully,

T. H. HUXLEY.

On December 16 he delivered a lecture "On the
Study of Biology," in connection with the Loan
Collection of Scientific Apparatus at South Kensington (*Coll. Essays*, iii. 262), dealing with the origin of
the name Biology, its relation to Sociology—"we
have allowed that province of Biology to become
autonomous ; but I should like you to recollect that
this is a sacrifice, and that you should not be
surprised if it occasionally happens that you see a

biologist apparently trespassing in the region of philosophy or politics; or meddling with human education; because, after all, that is a part of his kingdom which he has only voluntarily forsaken "— how to learn biology, the use of Museums, and above all, the utility of biology, as helping to give right ideas in this world, which "is after all, absolutely governed by ideas, and very often by the wildest and most hypothetical ideas."

This lecture on Biology was first published among the *American Addresses* in 1877.

It was about this time that an extremely Broad Church divine was endeavouring to obtain the signatures of men of science to a document he had drawn up protesting against certain orthodox doctrines. Huxley, however, refused to sign the protest, and wrote the following letter of explanation, a copy of which he sent to Mr. Darwin.

Nov. 18, 1876.

DEAR SIR—I have read the "Protest," with a copy of which you have favoured me, and as you wish that I should do so, I will trouble you with a brief statement of my reasons for my inability to sign it.

I object to clause 2 on the ground long since taken by Hume that the order of the universe such as we observe it to be, furnishes us with the only data upon which we can base any conclusion as to the character of the originator thereof.

As a matter of fact, men sin, and the consequences of their sins affect endless generations of their progeny. Men are tempted, men are punished for the sins of others without merit or demerit of their own ; and they are tormented for their evil deeds as long as their consciousness lasts.

The theological doctrines to which you refer, therefore, are simply extensions of generalisations as well based as any in physical science. Very likely they are illegitimate extensions of these generalisations, but that does not make them wrong in principle.

And I should consider it waste of time to "protest" against that which is.

As regards No. 3 I find that as a matter of experience, erroneous beliefs are punished, and right beliefs are rewarded—though very often the erroneous belief is based upon a more conscientious study of the facts than the right belief. I do not see why this should not be as true of theological beliefs as any others. And as I said before, I do not care to protest against that which is.

Many thanks for your congratulations. My tour was very pleasant and taught me a good deal.—I am yours very faithfully,   T. H. HUXLEY.

P.S.—You are at liberty to make what use you please of this letter.

4 MARLBOROUGH PLACE,
Nov. 19, 1876.

MY DEAR DARWIN—I confess I have less sympathy with the half-and-half sentimental school which he represents than I have with thoroughgoing orthodoxy.

If we are to assume that anybody has designedly set this wonderful universe going, it is perfectly clear to me that he is no more entirely benevolent and just in any intelligible sense of the words, than that he is malevolent and unjust. Infinite benevolence need not have invented pain and sorrow at all—infinite malevolence would very easily have deprived us of the large measure of content and happiness that falls to our lot. After all, Butler's "Analogy" is unassailable, and there is nothing in theological dogmas more contradictory to our moral sense, than is to be found in the facts of nature. From which, however, the Bishop's conclusion that the dogmas are true doesn't follow.—With best remembrances to Mrs. Darwin, ever yours very faithfully, T. H. HUXLEY.

This incident suggests the story of a retort he once made upon what he considered an unseasonable protest in church, a story which exemplifies, by the way, his strong sense of the decencies of life, appearing elsewhere in his constant respect for the ordinary conventions and his dislike for mere Bohemianism as such.

Once in a country house he was sitting at dinner next to his hostess, a lady who, as will sometimes happen, liked to play the part of Lady Arbitress of the whole neighbourhood. She told him how much she disapproved of the Athanasian Creed, and described how she had risen and left the village church when the parson began to read it ; and thinking to gain my father's assent, she turned to him and said graciously, " Now, Mr. Huxley, don't you think I was quite right to mark my disapproval ? "

" My dear Lady —— " he replied, " I should as soon think of rising and leaving your table because I disapproved of one of the entrées."

CHAPTER IX

1877

IN this year he delivered lectures and addresses on the "Geological History of Birds," at the Zoological Society's Gardens, June 7 ; on "Starfishes and their Allies," at the Royal Institution, March 7 ; at the London Institution, Dec. 17, on "Belemnites" (a subject on which he had written in 1864, and which was doubtless suggested anew by his autumn holiday at Whitby, where the Lias cliffs are full of these fossils); at the Anthropological Conference, May 22, on "Elementary Instruction in Physiology" (*Coll. Ess.* iii. 294), with special reference to the recent legislation as to experiments on living animals; and on "Technical Education" to the Working Men's Club and Institute, December 1 (*Coll. Ess.* iii. 404): a perilous subject, indeed, considering, as he remarks, that "any candid observer of the phenomena of modern society will readily admit that bores must be classed among the enemies of the human race ; and a little consideration will probably lead him to the further admission, that no species of that extensive genus of noxious creatures

is more objectionable than the educational bore.   . .
In the course of the last ten years, to go back no
farther, I am afraid to say how often I have ventured
to speak of education; indeed, the only part of this
wide region into which, as yet, I have not adventured, is that into which I propose to intrude to-
day."

The choice of subject for this address was connected
with a larger campaign for the establishment of
technical education on a proper footing, which began
with his work on the School Board, and was this year
brought prominently before the public by another
address delivered at the Society of Arts. The Cloth-
workers Company had already been assisting the
Society of Arts in their efforts for the spread of
technical education; and in July 1877 a special
committee of the Guilds applied to him, amongst
half a dozen others, to furnish them with a report as
to the objects and methods of a scheme of technical
education. This paper fills sixteen pages in the
Report of the Livery Companies' Committee for 1878.
The fundamental principles on which he bases his
practical recommendations are contained in the
following paragraph :—

It appears to me that if every person who is engaged
in an industry had access to instruction in the scientific
principles on which that industry is based ; in the mode
of applying these principles to practice; in the actual
use of the means and appliances employed ; in the
language of the people who know as much about the
matter as we do ourselves; and lastly, in the art of

keeping accounts, Technical Education would have done all that can be required of it.

And his suggestions about buildings was at once adopted by the Committee, namely, that they should be erected at a future date, regard being had primarily rather to what is wanted in the inside than what will look well from the outside.

Now the Guilds formed a very proper body to set such a scheme on foot, because only such wealthy and influential members of the first mercantile city in the world could afford to let themselves be despised and jeered at for professing to teach English manufacturers and English merchants that they needed to be taught; and to spend £25,000 a year towards that end for some time without apparent result.

That they eventually succeeded, is due no little to the careful plans drawn out by Huxley. He may be described as "really the engineer of the City and Guilds Institute; for without his advice," declared one of the leading members, "we should not have known what to have done."

At the same time he warned them against indiscriminate zeal; "though under-instruction is a bad thing, it is not impossible that over-instruction may be worse." The aim of the Livery Companies should specially be to aid the *practical* teaching of science, so that at bottom the question turns mainly on the supply of teachers.

On December 11, 1879, he found a further opportunity of urging the cause of Technical Education.

A lecture on Apprenticeships was delivered before the Society of Arts by Professor Silvanus Thompson. Speaking after the lecture (see report in *Nature*, 1879, p. 139) he discussed the necessity of supplying the place of the old apprenticeships by educating children in the principles of their particular crafts, beyond the time when they were forced to enter the workshops. This could be done by establishing schools in each centre of industry, connected with a central institution, such as was to be found in Paris or Zurich. As for complaints of deficient teaching of handicrafts in the Board Schools, it was more important for them to make intelligent men than skilled workmen, as again was indicated in the French system.

As President of the Royal Society, he was on the above-mentioned Committee of the Guilds from 1883 to 1885, and on December 10, 1883, distributed the prizes in connection with the institution in the Cloth-workers' Hall. After sketching the inception of the whole scheme, he referred to the Central Institute, then in course of building (begun in 1882, it was finished in 1884; the Technical College, Finsbury, was older by a year), and spoke of the difficulties in the way of organising such an institution :—

That building is simply the body, not the flesh and bones, but the bricks and stones, of the Central Institute, and the business upon which Sir F. Bramwell and my other colleagues on the Committee have been so much occupied, is the making a soul for this body; and I can assure you making a soul for anything is an amazingly difficult operation. You are always in danger of doing as the man in the story of Frankenstein did, and making something which will eventually devour you instead of being useful to you.

And here I may give a letter which refers to the movement for technical education, and the getting the City Companies under way in the matter. In the words of Mr. George Howell, M.P.,[1] it has an additional interest "as indicating the nature of his own epitaph"; as a man "whose highest ambition ever was to uplift the masses of the people and promote their welfare intellectually, socially, and industrially."

> 4 MARLBOROUGH PLACE, N.W.,
> *Jan.* 2, 1880.
>
> DEAR MR. HOWELL—Your letter is a welcome New Year's gift. There are two things I really care about—one is the progress of scientific thought, and the other is the bettering of the condition of the masses of the people by bettering them in the way of lifting themselves out of the misery which has hitherto been the lot of the majority of them. Posthumous fame is not particularly attractive to me, but, if I am to be remembered at all, I would rather it should be as "a man who did his best to help the people" than by other title. So you see it is no small pleasure and encouragement to me to find that I have been, and am, of any use in this direction.
>
> Ever since my experience on the School Board, I have been convinced that I should lose rather than gain by entering directly into politics. . . . But I suppose I have

---

[1] Who sent it to the *Times* (July 3, 1895) just after Huxley's death.

some ten years of activity left in me, and you may
depend upon it I shall lose no chance of striking a blow
for the cause I have at heart. I thought the time had
come the other day at the Society of Arts, and the event
proves I was not mistaken. The animal is moving, and
by a judicious exhibition of carrots in front and kicks
behind, we shall get him into a fine trot presently. In
the meantime do not let the matter rest. . . . The (City)
companies should be constantly reminded that a storm is
brewing. There are excellent men among them, who
want to do what is right, and need help against the
sluggards and reactionaries. It will be best for me to
be quiet for a while, but you will understand that I am
watching for the turn of events.—I am, yours very faithfully,   T. H. HUXLEY.

This summer, too, he delivered a course on Biology
for Teachers at South Kensington, and published
not only his *American Addresses*, but also the *Physiography*, founded upon the course delivered seven
years before. The book, of which 3386 copies were
sold in the first six weeks, was fruitful in two ways ;
it showed that a geographical subject could be
invested with interest, and it set going what was
almost a new branch of teaching in natural science,
even in Germany, the starting place of most educational methods, where it was immediately proposed
to bring out an adaptation of the book, substituting,
e.y. the Elbe for the Thames, as a familiar example of
river action.

He was immensely pleased by a letter from Mr.
John Morley, telling how his step-son, a boy of non-
bookish tastes, had been taken with it. "My step-

son was reading it the other night. I said, 'Isn't it better to read a novel before going to bed, instead of worrying your head over a serious book like that?' 'Oh,' said he, 'I'm at an awfully interesting part, and I can't leave off.'" It was, Mr. Morley continued, "the way of making Nature, as she comes before us every day, interesting and intelligible to young folks."

To this he replied on December 14 :—

I shall get as vain as a peacock if discreet folk like you say such pretty things to me as you do about the *Physiography*.

But it is very pleasant to me to find that I have succeeded in what I tried to do. I gave the lectures years ago to show what I thought was the right way to lead young people to the study of nature—but nobody would follow suit—so now I have tried what the book will do.

Your step-son is a boy of sense, and I hope he may be taken as a type of the British public !

A good deal of time was taken up in the first half of the year by the Scottish Universities Commission, which necessitated his attendance in Edinburgh the last week in February, the first week in April, and the last week in July. He had hoped to finish off the necessary business at the first of these meetings, but no sooner had he arrived in Edinburgh, after a pleasant journey down with J. A. Froude, than he learned that "the chief witness we were to have examined to-day, and whose due evisceration was one of the objects of my coming, has telegraphed to say he can't be here." Owing to this and to the enforced absence of the judges on the Commission from some of the sittings, it was found necessary to have additional meetings at Easter, much to his disgust. He writes :—

> I am sorry to say I shall have to come here again in Easter week. It is the only time the Lord President is free from his courts, and although we all howled privately, there was no help for it. Whether we finish then or not will depend on the decision of the Government, as to our taking up the case of you troublesome women, who want admission into the University (very rightly too I think). If we have to go into this question it will involve the taking of new evidence and no end of bother. I find my colleagues very reasonable, and I hope some good may be done, that is the only consolation.
>
> I went out with Blackie last evening to dine with the Skeltons, at a pretty place called the Hermitage, about three miles from here. . . . Blackie and I walked home with snow on the ground and a sharp frost. I told you it would turn cold as soon as I got here, but I am none the worse.

It was just the same in April :—

> It is quite cold here as usual, and there was ice on the ponds we passed this morning. . . . I am much better lodged than I was last time, for the same thanks to John Bruce, but I do believe that the Edinburgh houses are the coldest in the universe. In spite of a good breakfast and a good fire, the half of me that is writing to you is as cold as charity.
>
> April 4.—We toil at the Commission every day, and don't make any rapid progress. An awful fear creeps over me that we shall not finish this bout.

While he was in Edinburgh for the third time, his attention was called to an article in the *Echo*, the organ of the anti-vivisection party.   He writes :—

The *Echo* is pretty. It is one of a long series of articles from the same hand, but I don't think they hurt anybody and they evidently please the writer. For some reason or other they have not attacked me yet, but I suppose my turn will come.

Again :—

Thank you for sending me John Bright's speeches. They are very good, but hardly up to his old mark of eloquence. Some parts are very touching.

His health was improving, as he notes with satisfaction :

Every day this week we have had about four hours of the Commission, and I have dined out four days out of the six. But I'm no the waur, and the late dinners have not been visited by fits of morning blue devils. So I am in hopes that I am getting back to the normal state that Clark prophesied for me.

4 MARLBOROUGH PLACE, LONDON, N.W.,
April 29, 1877.

MY DEAR SKELTON—Best thanks for your second edition. You paint the system [1] in such favourable colours, that I am thinking of taking advantage of it for my horde of "young barbarians." I am sure Scotch air would be of service to them—and in after-life they might have the inestimable advantage of a quasi-Scotch nationality—that greatest of all practical advantages in Britain.

---

[1] *I.e.* of Scotch education.

We are to sit again in the end of July when Mrs.
Skelton and you, if you are wise, will be making holiday.

Your invitation is most tempting, and if I had no
work to do I should jump at it.

But alas! I shall have a deal of work, and I must
go to my Patmos in George Street. Ingrained laziness
is the bane of my existence; and you don't suppose that
with the sun shining down into your bosky dell, and
Mrs. Skelton radiant, and Froude and yourself nicotiant,
I am such a Philistine as to do a stroke of work?—Ever
yours very faithfully,

T. H. HUXLEY.

From Edinburgh he went to St. Andrews to make
arrangements for his elder son to go to the University
there as a student the following winter. Then he
paid a visit to Sir W. Armstrong in Northumberland,
afterwards spending a month at Whitby. His holiday
work consisted in a great part of the article on
"Evolution" for the *Encyclopædia Britannica*, which
is noted as finished on October 24, though not
published till the next year.

In November the honorary degree of LL.D. was
conferred upon Charles Darwin at Cambridge, "a
great step for Cambridge, though it may not seem
much in itself," he writes to Dohrn, November 21.
In the evening after the public ceremony there was
a dinner of the Philosophical Club, at which he spoke
in praise of Darwin's services to science. Darwin
himself was unable to be present, but received an
enthusiastic account of the proceedings from his son,
and wrote to thank Huxley, who replied :—

4 MARLBOROUGH PLACE,
Nov. 21, 1877.

MY DEAR DARWIN—Nothing ever gave me greater pleasure than the using the chance of speaking my mind about you and your work which was afforded me at the dinner the other night. I said not a word beyond what I believe to be strictly accurate : and, please Sir, I didn't sneer at anybody. There was only a little touch of the whip at starting, and it was so tied round with ribbons that it took them some time to find out where the flick had hit.   T. H. HUXLEY.

He writes to his wife :—

I will see if I can recollect the speech. I made a few notes sitting in Dewar's room before the dinner. But as usual I did not say some things I meant to say, and said others that came up on the spur of the moment.

And again :—

Please I didn't say that Réaumur was the other greatest scientific man since Aristotle. But I said that in a certain character of his work he was the biggest man between Aristotle and Darwin. I really must write out an "authorised version" of my speech. I hear the Latin oration is to be in *Nature* this week, and Lockyer wanted me to give him the heads of my speech, but I did not think it would be proper to do so, and refused. I have written out my speech as well as I can recollect it. I do not mind any friend seeing it, but you must not let it get about as the dinner was a private one.

The notes of his speech run as follows :—

MR. PRESIDENT—I rise with pleasure and with alacrity to respond to the toast which you have just proposed, and I may say that I consider one of the greatest honours

which have befallen me, to be called upon to represent
my distinguished friend Mr. Darwin upon this occasion.
I say to represent Mr. Darwin, for I cannot hope to
personate him, or to say all that would be dictated by
a mind conspicuous for its powerful humility and strong
gentleness.

Mr. Darwin's work had fully earned the distinction
you have to-day conferred upon him four-and-twenty
years ago; but I doubt not that he would have found
in that circumstance an exemplification of the wise fore-
sight of his revered intellectual mother.   Instead of
offering her honours when they ran a chance of being
crushed beneath the accumulated marks of approbation
of the whole civilised world, the University has waited
until the trophy was finished, and has crowned the edifice
with the delicate wreath of academic appreciation.

This is what I suppose Mr. Darwin might have said
had he been happily able to occupy my place.   Let me
now speak in my own person and in obedience to your
suggestion, let me state as briefly as possible what appear
to me to be Mr. Darwin's distinctive merits.

From the time of Aristotle to the present day I know
of but one man who has shown himself Mr. Darwin's
equal in one field of research—and that is Réaumur.   In
the breadth of range of Mr. Darwin's investigations upon
the ways and works of animals and plants, in the minute
patient accuracy of his observations, and in the philosophical ideas which have guided them, I know of no
one who is to be placed in the same rank with him
except Réaumur.

Secondly, looking back through the same long period
of scientific history, I know of but one man, Lyonnet,
who not being from his youth a trained anatomist, has
published such an admirable minute anatomical research
as is contained in Mr. Darwin's work on the Cirripedes.

Thirdly, in that region which lies between Geology
and Biology, and is occupied by the problem of the

influence of life on the structure of the globe, no one, so far as I know, has done a more brilliant and far-reaching piece of work than the famous book upon Coral Reefs.

I add to these as incidental trifles the numerous papers on Geology, and that most delightful of popular scientific books, the *Journal of a Naturalist*, and I think I have made out my case for the justification of to-day's proceedings.

But I have omitted something. There is the *Origin of Species*, and all that has followed it from the same marvellously fertile brain.

Most people know Mr. Darwin only as the author of this work, and of the form of evolutional doctrine which it advocates. I desire to say nothing about that doctrine. My friend Dr. Humphry has said that the University has by to-day's proceedings committed itself to the doctrine of evolution. I can only say "I am very glad to hear it." But whether that doctrine be true or whether it be false, I wish to express the deliberate opinion, that from Aristotle's great summary of the Biological knowledge of his time down to the present day, there is nothing comparable to the *Origin of Species*, as a connected survey of the phenomena of life permeated and vivified by a central idea. In remote ages the historian of science will dwell upon it as the starting-point of the Biology of his present and our future.

My friend Dr. Humphry has adverted to somebody about whom I know nothing, who says that the exact and critical studies pursued in this University are ill-calculated to preserve a high tone of mind.

I presume that this saying must proceed from some one wholly unacquainted with Cambridge. Whoever he may be, I beg him, if he can, to make the acquaintance of Charles Darwin.

In Mr. Darwin's name I beg leave to thank you for the honour you have done him.

It happened that the quadrennial election of a Lord Rector at St. Andrews University fell in this year, and on behalf of a number of students, Huxley received a telegram from his son, now newly entered at St. Andrews, asking him to stand. He writes to his wife :—

That boy of yours has just sent me a telegram, which I enclose. I sent back message to say that as a Commissioner on the Scotch Universities I could not possibly stand. The cockerel is beginning to crow early. I do believe that to please the boy I should have assented to it if it had not been for the R. Commission.

Apropos of controversies (November 23)

We had a grand discussion at the Royal Society last night between Tyndall and Burdon Sanderson. The place was crammed, and we had a late sitting. I'm not sure, however, that we had got much further at the end than at the beginning, which is a way controversies have.

The following story is worth recording, as an illustration not only of the way in which Huxley would give what help was in his power to another man of science in distress, but of the ready aid proffered on this, as on many other occasions, by a wealthy northern merchant who was interested in science. A German scientific worker in England, whom we will call H., had fallen into distress, and applied to him for help, asking if some work could not be put in his way. Huxley could think of nothing immediate but to suggest some lessons in German literature to his children, though in fact

they were well provided for with a German governess; nevertheless he thought it a proper occasion
to avail himself of his friend's offer to give help in
deserving cases.   He writes to his wife :—

I made up my mind to write to X. the day before
yesterday; this morning by return of post he sends me
a cheque not only for the £60 which I said H. needed,
but £5 over for his present needs with a charming letter.

It came in the nick of time, as H. came an hour or
two after it arrived, and with many apologies told me
he was quite penniless.   The poor old fellow was quite
overcome when I told him of how matters stood, and it
was characteristic that as soon as he got his breath again,
he wanted to know when he would begin teaching the
children !   I sent him to get an order on the Naples
bank for discharge of his debt there.   X.'s express
stipulation was that his name should not be mentioned,
so mind you say not a word about his most kind and
generous act.

The following letters of miscellaneous interest
were written in this year :—

4 MARLBOROUGH PLACE,
Nov. 21, 1877

MY DEAR MORLEY—I am always at the command of
the *Fortnightly* so long as you are editor, but I don't think
that the Belemnite [1] business would do for you.   The
story would hardly be intelligible without illustrations.

There are two things I am going to do which may be
more to the purpose.   One is a screed on Technical
Education which I am going to give to the Working
Men's Union on the 1st December.

The other is a sort of Éloge on Harvey at the Royal

---

[1] The lecture at the London Institution mentioned above.

Institution in March apropos of his 300th birthday—which was Allfools Day.

You shall have either of these you like, but I advise Harvey ; as if I succeed in doing what I shall aim at it will be interesting.

Why the deuce do you live at Brighton ? St. John's Wood is far less cockneyfied, and its fine and Alpine air would be much better for you, and I believe for Mrs. Morley, than the atmosphere of the melancholy main, the effects of which on the human constitution have been so well expounded by that eminent empiric, Dr. Dizzy.

Anyhow, I wish we could see something of you now and then.—Ever yours very faithfully,

T. H. HUXLEY.

Darwin got his degree with great *éclat* on Saturday. I had to return thanks for his health at the dinner of the Philosophical Society ; and oh ! I chaffed the dons so sweetly.

4 MARLBOROUGH PLACE, N.W.,
Nov. 27, 1877.

MY DEAR MORLEY—You shall have both the articles—if it is only that I may enjoy the innocent pleasure of Knowles' face [1] when I let him know what has become of them.

Stormy ocean, forsooth ! I back the storm and rain through which I came home to-night against anything London-super-mare has to show.

I will send the MS. to Virtue as soon as it is in a reasonable state.—Ever yours very faithfully,

T. H. HUXLEY.

4 MARLBOROUGH PLACE, N.W.,
Jan. 8, 1878.

MY DEAR MORLEY—Many thanks for the cheque. In my humble judgment it is quite as much as the commodity is worth.

---

[1] The rival editor. Cp. p. 150 above.

It was a great pleasure to us all to have you with us on New Year's Day. My wife claims it as her day, and I am not supposed to know anything about the guests except Spencer and Tyndall. None but the very elect are invited to the sacred feast—so you see where you stand among the predestined who cannot fall away from the state of grace.

I have not seen Spencer in such good form and good humour combined for an age.

I am working away at Harvey, and will send the MS. to Virtue's as soon as I am sufficiently forward.—Ever yours very faithfully,   T. H. HUXLEY.

4 MARLBOROUGH PLACE,
Dec. 9, 1877.

MY DEAR TYNDALL—I am so sorry to have been out when Mrs. Tyndall called to-day. By what we heard at the *x* on Thursday, I imagined you were practically all right again, or I should have been able to look after you to-day.

But what I bother you with this note for is to beg you not to lecture at the London Institution to-morrow, but to let me change days with you, and so give yourself a week to recover. And if you are seedy, then I am quite ready to give them another lecture on the Hokypotamus or whatever else may turn up.

But don't go and exert yourself in your present condition. These severe colds have often nothing very tangible about them, but are not to be trifled with when folks are past fifty.

Let me have an answer to say that I may send a telegram to Nicholson first thing to-morrow morning to say that I will lecture *vice* you. My "bottled life," as Hutton calls it in the *Spectator* [1] this week, is quite ready to go off.

---

[1] The *Spectator* for Dec. 8, 1877, began an article thus:—
" Professor Huxley delivered a very amusing address last Saturday

Now be a sane man and take my advice.—Ever yours
very faithfully,   T. H. HUXLEY.

---

at the Society of Arts, on the very unpromising subject of technical
education ; but we believe that if Professor Huxley were to become
the President of the Social Science Association, or of the International Statistical Congress, he would still be amusing, so much
bottled life does he infuse into the driest topic on which human
beings ever contrived to prose."

CHAPTER X

1878

THE year 1878 was the tercentenary of Harvey's birth, and Huxley was very busy with the life and work of that great physician. He spoke at the memorial meeting at the College of Physicians (July 18), he gave a lecture on Harvey at the Royal Institution on January 25, afterwards published in *Nature* and the *Fortnightly Review*, and intended to write a book on him in a projected *English Men of Science* series (see p. 255 sq. *infra*).

I am very glad you like "Harvey" (he writes to Prof. Baynes on Feb. 11). He is one of the biggest scientific minds we have had. I expect to get well vilipended not only by the anti-vivisection folk, for the most of whom I have a hearty contempt, but *apropos* of Bacon. I have been oppressed by the humbug of the "Baconian Induction" all my life, and at last *the worm has turned.*

Now in this lecture he showed that Harvey employed vivisection to establish the doctrine of the circulation of the blood, and furthermore, that he taught this doctrine before the *Novum Organum* was published, and that his subsequent *Exercitatio* displays no trace of being influenced by Bacon's work. After glancing at the superstitious reverence for the "Baconian Induction," he pointed out Bacon's ignorance of the progress of science up to his time, and his inability to divine the importance of what he knew by hearsay of the work of Copernicus, or Kepler, or Galileo; of Gilbert, his contemporary, or of Galen; and wound up by quoting Ellis's severe judgment of Bacon in the General Preface to the Philosophic Works, in Spedding's classical edition (p. 38):—"That his method is impracticable cannot, I think, be denied, if we reflect, not only that it never has produced any result, but also that the process by which scientific truths have been established cannot be so presented as even to appear to be in accordance with it."

How early this conviction had forced itself upon him, I cannot say; but it was certainly not later than 1859, when the *Origin of Species* was constantly met with "Oh, but this is contrary to the Baconian method." He had long felt what he expresses most clearly in the "Progress of Science" (*Coll. Ess.* i. 46-57), that Bacon's "majestic eloquence and fervid vaticinations," which "drew the attention of all the world to the 'new birth of Time,'" were yet, for all practical results on discovery, "a magnificent failure." The desire for "fruits" has not been the great motive of the discoverer; nor has discovery waited upon collective research.   "Those who refuse to go beyond

fact," he writes, "rarely get as far as fact; and any
one who has studied the history of science knows
that almost every great step therein has been made
by the 'anticipation of nature,' that is, by the in-
vention of hypotheses, which, though verifiable, often
had very little foundation to start with; and, not
unfrequently, in spite of a long career of usefulness,
turned out to be wholly erroneous in the long-
run."

Thus he had been led to a settled disbelief in
Bacon's scientific greatness, that reasoned "prejudice"
against which Spedding himself was moved to write
twice in defence of Bacon. In his first letter he
criticised a passage in the lecture touching this
question. On the one hand, he remarks, "Bacon
would probably have agreed with you as to his pre-
tensions as a scientific discoverer (he calls himself a
bellman to call other wits together, or a trumpeter,
or a maker of bricks for others to build with)." On
the other hand, he asks, ought a passage from a
fragment—the *Temporis partus masculus*—unpublished
in Bacon's lifetime, to be treated as one of his representative opinions?

In his second letter he adduces, on other grounds,
his own more favourable impression of Bacon's philosophical influence. A peculiar interest of this letter
lies in its testimony to the influence of Huxley's
writings even on his elder contemporaries.

From James Spedding

Feb. 1, 1878.

When you admit that you study Bacon with a *prejudice*, you mean of course an unfavourable opinion previously formed on sufficient grounds. Now I am myself supposed to have studied him with a prejudice the other way : but this I cannot admit, in any sense of the word ; for when I first made his acquaintance I had no opinion or feeling about him at all—more than the ordinary expectation of a young man to find what he is told to look for. My earliest impression of his character came probably from Thomson—whose portrait of him, except as touched and softened by the tenderer hand of "the sweet-souled poet of the Seasons," did not differ from the ordinary one. It was not long indeed before I did begin to form an opinion of my own ; one of those *after*-judgments which are liable to be mistaken for prejudices by those who judge differently, and which, being formed, do, no doubt, tell upon the balance. For it was not long before I found myself indebted to him for the greatest benefit probably that any man, living or dead, can confer on another. In my school and college days I had been betrayed by an ambition to excel in themes and declamations into the study, admiration, and imitation of the rhetoricians. In the course of my last long vacation—the autumn of 1830—I was inspired with a new ambition, namely, to think justly about everything which I thought about at all, and to act accordingly ; a conviction for which I cannot cease to feel grateful, and which I distinctly trace to the accident of having in the beginning of that same vacation given two shillings at a second-hand bookstall for a little volume of Dove's classics, containing the Advancement of Learning. And if I could tell you how many superlatives I have since that time degraded into the positive ; how many innumerables and infinites I have replaced by counted numbers and estimated quantities; how many assumptions, important to the argument in hand, I have withdrawn because I found on more consideration that the fact might be explained otherwise; and how many effective epithets I have discarded when I found that I could not fully verify them; you would think it no less than just that I should claim for myself and concede to others the right of being judged by the last edition rather than the first. That a persistent endeavour to free myself from what you regard as Bacon's characteristic vice should have been the fruit of a desire to follow his example, will seem strange to you, but it is fact. Perhaps you will think it not less strange, but it is my real belief, that if your own writings had been in existence and come in my way at the same critical stage of my moral and mental development, they would have taught me the same lesson and inspired me with the same ambition; for in that particular (if I may say it without offence) I look upon you *both* as eminent examples of the *same* virtue.

To the lecture he refers once more in a letter to Mr. John Morley. The political situation touched on in this and the next letter is that of the end of the Russo-Turkish war and the beginning of the Afghan war.

SCIENCE SCHOOLS, SOUTH KENSINGTON,
Feb. 7, 1878.

MY DEAR MORLEY—Many thanks for the cheque, and still more for your good word for the article.[1] I knew it would "draw" Hutton, and his ingenuity has as usual made the best of the possibilities of attack. I am

---

[1] On Harvey.

glad to find, however, that he does not think it expedient to reiterate his old story about the valuelessness of vivisection in the establishment of the doctrine of the circulation.

I hear that that absurd creature R—— goes about declaring that I have made all sorts of blunders. Could not somebody be got to persuade him to put what he has to say in black and white?

Controversy is as abhorrent to me as gin to a reclaimed drunkard; but oh dear! it would be so nice to squelch that pompous impostor.

I hope you admire the late aspects of the British Lion. His tail goes up and down from the intercrural to the stiffly erect attitude per telegram, while his head is sunk in the windbag of the House of Commons.

I am beginning to think that a war would be a good thing if only for the inevitable clean sweep of all the present governing people which it would bring about.— Ever yours very faithfully,   T. H. HUXLEY.

To his Eldest Daughter

SCIENCE SCHOOLS, SOUTH KENSINGTON,
Dec. 7, 1878.

DEAREST JESS—You are a badly used young person —you are; and nothing short of that conviction would get a letter out of your still worse used Pater, the *bête noire* of whose existence is letter-writing.

Catch me discussing the Afghan question with you, you little pepper pot. No, not if I know it. Read Fitzjames Stephen's letter in the *Times*, also Bartle Frere's memorandum, also Napier of Magdala's memo. Them's my sentiments.

Also read the speech of Lord Hartington on the address. He is a man of sense like his father, and you will observe that he declares that the Government were perfectly within their right in declaring war without calling Parliament together. . . .

If you had lived as long as I have and seen as much of men, you would cease to be surprised at the reputations men of essentially commonplace powers—aided by circumstances and some amount of cleverness—obtain.

I am as strong for justice as any one can be, but it is real justice, not sham conventional justice which the sentimentalists howl for.

At this present time real justice requires that the power of England should be used to maintain order and introduce civilisation wherever that power extends.

The Afghans are a pack of disorderly treacherous blood-thirsty thieves and caterans who should never have been allowed to escape from the heavy hand we laid upon them, after the massacre of twenty thousand of our men, women (and) children in the Khoord Cabul Pass thirty years ago.

We have let them be, and the consequence is they now lend themselves to the Russians, and are ready to stir up disorder and undo all the good we have been doing in India for the last generation.

They are to India exactly what the Highlanders of Scotland were to the Lowlanders before 1745 ; and we have just as much right to deal with them in the same way.

I am of opinion that our Indian Empire is a curse to us. But so long as we make up our minds to hold it, we must also make up our minds to do those things which are needful to hold it effectually, and in the long-run it will be found that so doing is real justice both for ourselves, our subject population, and the Afghans themselves.

There, you plague.—Ever your affec. Daddy,

T. H. HUXLEY.

A few days later he writes to his son :—

The Liberals are making fools of themselves, and " the
family " declare I am becoming a Jingo ! Another speech
from Gladstone is expected to complete my conversion.

Among other occupations he still had to attend
the Scottish Universities Commission, for which he
wrote the paragraph on examinations in its report;
he lectured on the Hand at the Working Men's
College; prepared new editions of the *Physiography,
Elementary Physiology,* and *Vertebrate Anatomy,* and at
length brought out the *Introductory Primer* in the
Science Primer Series, in quite a different form from
what he had originally sketched out. But his chief
interest lay in the Invertebrata. From April 29 to
June 3 he lectured to working men at Jermyn Street
upon the Crayfish; read a paper on the Classification
and Distribution of Crayfishes at the Zoological
Society on June 4, and lectured at the Zoological
Gardens weekly from May 17 to June 21 on Crus-
taceous Animals. In all this work lay the foundations
of his subsequent book on the Crayfish, which I find
jotted down in the notes of this year to be written
as an introduction to *Zoology,* together with the
"Dog" as an introduction to the *Mammalia,* and
Man—already dealt with in *Man's Place in Nature*—
as an introduction to *Anthropology.* This projected
series is completed with a half-erased note of an
introduction to *Psychology,* which perhaps found some
expression in parts of the *Hume,* also written this
year.

He notes down also, work on the Ascidians, and

on the morphology of the Mollusca and Cephalopods brought back by the *Challenger*, in connection with which he now began the monograph on the rare creature Spirula, a remarkable piece of work, being based upon the dissections of a single specimen, but destined never to be completed by his hand, though his drawings were actually engraved, and nothing remained but to put a few finishing touches and to write detailed descriptions of the plates.

Letters to W. K. Parker and Professor Haeckel touch on this part of his work; the former, indeed, offering a close parallel to a story, obviously of the same period, which the younger Parker tells in his reminiscences, to illustrate the way in which he would be utterly engrossed in a subject for the time being. Jeffery Parker, while demonstrator of biology, came to him with a question about the brain of the codfish at a time when he was deep in the investigation of some invertebrate group. "Codfish?" he replied, "that's a vertebrate, isn't it? Ask me a fortnight hence, and I'll consider it."

4 MARLBOROUGH PLACE,
Sept. 25, 1878.

MY DEAR PARKER—As far as I recollect *Ammocœtes* is a vertebrated animal—and I ignore it.

The paper you refer to was written by my best friend —a carefulish kind of man—and I am as sure that he saw what he says he saw, as if I had seen it myself.

But what the fact may mean and whether it is temporary or permanent—is thy servant a dog that he should worry himself about other things with backbones ? Not if I know it.

Churchill has got over a whole batch of the American edition of the *Vertebrata*, so I have a respite. Mollusks are far more interesting—bugs sweeter—while the dinner crayfish hath no parallel for intense and absorbing interest in the three kingdoms of Nature.

What saith the Scripture ? "Go to the ANT, thou sluggard." In other words, study the Invertebrata.— Ever yours very faithfully,   T. H. HUXLEY.

[Sketch of a vast winged ant advancing on a midget, and saying, as it looks through a pair of eyeglasses, "Well, really, what an absurd creature ! !"]

4 MARLBOROUGH PLACE, LONDON,
April 28, 1878.

MY DEAR HAECKEL—Since the receipt of your letter three months ago, I have been making many inquiries about *Medusæ* for you, but I could hear of none—and so I have delayed my reply, until I doubt not you have been blaspheming my apparent neglect.

My "Sammlung"!!¹ My dear friend, my cabin on board H.M.S. *Rattlesnake* was 7 feet long, 6 feet wide, and 5 feet 6 inches high. When my bed and my clothes were in it, there was not much room for any collection, except the voluntary one made by some thousands of specimens of *Blatta Orientalis*,² with whose presence I should have been very glad to dispense.

My *Medusæ* were never published. I have heaps of notes and drawings and half-a-dozen engraved plates. But after the publication of the *Oceanic Hydrozoa* I was obliged to take to quite other occupations, and all that material is like the "full many a flower, born to blush unseen," of our poet.

---

¹ Collection.   ² The cockroach.

If you would pay us a visit you should look through the whole mass, if you liked, and you might find something interesting.

At present, I am very busy about Crayfishes (Flusskrebse), working out the relations between their structure and their Geographical Distribution, which are very curious and interesting.

I have also nearly finished the anatomy of *Spirula* for the *Challenger*. It is essentially a cuttlefish, and the shell is really internal. With only one specimen, it has been a long and troublesome job—but I shall establish all the essential points and give half-a-dozen plates of anatomy.

You will recollect my eldest little daughter? She is going to be married next Saturday. It is the first break in our family, and we are very sad to lose her—though well satisfied with her prospects. She is but just twenty and a charming girl, though you may put that down to fatherly partiality if you like.

The second daughter has taken to art, and will make a painter if she be wise enough not to marry for some years.

My eldest son who comes next is taller than I am. He has been at one of the Scotch Universities for the last six months; and one of these fine days, next month, you will see a fair-haired stripling asking for Herr Professor Haeckel.

I am going to send him to Jena for three months to pick up your noble vernacular; and in the meanwhile to continue his Greek and Mathematics, in which the young gentleman is fairly proficient. If you can recommend any Professor under whom he can carry on his studies, it will be a great kindness.

I will give him a letter to you, and while I beg you not to give yourself any trouble about him, I need not say I shall be very grateful for any notice you may take of him.

I am giving him as much independence of action as possible, in order that he may learn to take care of himself.

Now that is enough about my children. Yours must yet be young—and you have not yet got to the marriage and university stage—which I assure you is much more troublesome than the measles and chicken-pox period.

My wife unites with me in kindest remembrances and good wishes.—Ever yours very faithfully,

T. H. HUXLEY.

An outbreak of diphtheria among his children made the spring of 1878 a time of overwhelming anxiety. How it told upon his strong and self-contained chief is related by T. J. Parker—"I never saw a man more crushed than he was during the dangerous illness of one of his daughters, and he told me that, having then to make an after-dinner speech, he broke down for the first time in his life, and for one painful moment forgot where he was and what he had to say." This was one of the few occasions of his absence from College during the seventies. "When, after two days, he looked in at the laboratory," writes Professor Howes, "his dejected countenance and tired expression betokened only too plainly the intense anxiety he had undergone."

The history of the outbreak was very instructive. Huxley took a leading part in organising an inquiry and in looking into the matter with the health officer. "As soon as I can get all the facts together," he writes on Dec. 10, "I am going to make a great turmoil about our outbreak of diphtheria—and see whether I cannot get our happy-go-lucky local government

mended." As usual, the epidemic was due to culpable
negligence. In the construction of some drains, too
small a pipe was laid down. The sewage could not
escape, and flooded back in a low-lying part of
Kilburn. Diphtheria soon broke out close by.
While it was raging there, a St. John's Wood
dairyman running short of milk, sent for more to an
infected dairy in Kilburn. Every house which he
supplied that day with Kilburn milk was attacked
with diphtheria.

But with relief from this heavy strain, his spirits
instantly revived, and he writes to Tyndall.

> 4 MARLBOROUGH PLACE,
> *May* 20, 1878.
>
> MY DEAR TYNDALL—I wrote you a most downhearted
> letter this morning about Madge, and not without reason.
> But having been away four hours, I come home to find
> a wonderful and blessed change. The fever has abated
> and she is looking like herself. If she could only make
> herself heard, I should have some sauciness. I see it in
> her eyes.
>
> If you will be so kind as to kiss everybody you meet
> on my account it will be a satisfaction to me. You may
> begin with Mrs. Tyndall!—Ever yours,   T. H. HUXLEY.

Professor Marsh, with whom Huxley had stayed
at Yale College in 1876, paid his promised visit to
England immediately after this.

> 4 MARLBOROUGH PLACE, N.W.,
> *June* 24, 1878. (*Evening*).
>
> MY DEAR MARSH—Welcome to England! I am
> delighted to hear of your arrival—but the news has only

just reached me, as I have been away since Saturday
with my wife and sick daughter who are at the seaside.
A great deal has happened to us in the last six or seven
weeks. My eldest daughter married, and then a week
after an invasion of diphtheria, which struck down my
eldest son, my youngest daughter, and my eldest remaining
daughter all together. Two of the cases were light, but
my poor Madge suffered terribly, and for some ten days
we were in sickening. anxiety about her. She is slowly
gaining strength now, and I hope there is no more cause
for alarm—but my household is all to pieces—the Lares
and Penates gone, and painters and disinfectors in their
places.

You will certainly have to run down to Margate and
see my wife—or never expect forgiveness in this world.

I shall be at the Science Schools, South Kensington,
to-morrow till four—and if I do not see you before that
time I shall come and look you up at the Palace Hotel.—
I am, yours very faithfully,   T. H. HUXLEY.

"Is it not provoking," he writes to his wife, "that
we should all be dislocated when I should have been
so glad to show him a little attention?" Still, apart
from this week-end at the seaside, Professor Marsh
was not entirely neglected. He writes in his *Recollections* (p. 6) :—

How kind Huxley was to every one who could claim
his friendship, I have good cause to know. Of the many
instances which occur to me, one will suffice. One evening
in London at a grand annual reception of the Royal
Academy, where celebrities of every rank were present,
Huxley said to me, "When I was in America, you showed
me every extinct animal that I had read about, or even
dreamt of. Now, if there is a single living lion in all
Great Britain that you wish to see, I will show him to

you in five minutes." He kept his promise, and before the reception was over, I had met many of the most noted men in England, and from that evening, I can date a large number of acquaintances, who have made my subsequent visits to that country an ever-increasing pleasure.

As for his summer occupations, he writes to his eldest daughter on July 2 :—

No, young woman, you don't catch me attending any congresses I can avoid, not even if F. is an artful committee-man. I must go to the British Association at Dublin—for my sins—and after that we have promised to pay a visit in Ireland to Sir Victor Brooke. After that I must settle myself down in Penmaenmawr and write a little book about David Hume—before the grindery of the winter begins.

The meeting of the British Association took place this year in the third week of August at Dublin. Huxley gave an address in the Anthropological subsection,[1] and on the 20th received the honorary degree of LL.D. from Dublin University, the Public Orator presenting him in the following words :—

Præsento vobis Thomam Henricum Huxley—hominem vere physicum—hominem facundum, lepidum, venustum —eundem autem nihil (philosophia modo sua lucem præferat) reformidantem—ne illud quidem Ennianum,
Simia quam similis, turpissima bestia, nobis.

The extract above given contains the first reference to the book on Hume,[2] written this summer as

[1] "Informal Remarks on the Conclusions of Anthropology," *B. A. Report*, 1878, pp. 573-578.
[2] In the "English Men of Letters" series, edited by Mr. John Morley.

a holiday occupation at Penmaenmawr. The speed at which it was composed is remarkable, even allowing for his close knowledge of the subject, acquired many years before. Though he had been "picking at it" earlier in the summer, the whole of the philosophical part was written during September, leaving the biographical part to be done later.

The following letters from Marlborough Place show him at work upon the book :—

March 31, 1878.

My DEAR MORLEY—I like the notion of undertaking your Hume book, and I don't see why I should not get it done this autumn. But you must not consider me pledged on that point, as I cannot quite command my time.

Tulloch sent me his book on Pascal. It was interesting as everything about Pascal must be, but Tulloch is not a model of style.

I have looked into Bruton's book, but I shall now get it and study it. Hume's correspondence with Rousseau seems to me typical of the man's sweet, easy-going nature. Do you mean to have a portrait of each of your men? I think it is a great comfort in a biography to get a notion of the subject in the flesh.

I have rather made it a rule not to part with my property in my books—but I daresay that can be arranged with Macmillan. Anyhow I shall be content to abide by the general arrangement if you have made one.

We have had a bad evening. Clifford [1] has been here, and he is extremely ill—in fact I fear the worst for him.

It is a thousand pities, for he has a fine nature all round, and time would have ripened him into something very considerable. We are all very fond of him.—Ever yours very faithfully,   T. H. HUXLEY.

[1] See p. 259.

July 6, 1878.

MY DEAR MORLEY—Very many thanks for Diderot. I have made a plunge into the first volume and found it very interesting.   I wish you had put a portrait of him as a frontispiece.   I have seen one—a wonderful face, something like Goethe's.

I am picking at Hume at odd times.   It seems to me that I had better make an analysis and criticism of the " Inquiry," the backbone of the essay—as it touches all the problems which interest us most just now.   I have already sketched out a chapter on Miracles, which will, I hope, be very edifying in consequence of its entire agree- ment with the orthodox arguments against Hume's *a priori* reasonings against miracles.

Hume wasn't half a sceptic after all.   And so long as he got deep enough to worry Orthodoxy, he did not care to go to the bottom of things.

He failed to see the importance of suggestions already made both by Locke and Berkeley.—Ever yours very faithfully,   T. H. HUXLEY.

Sept. 30, 1878.

MY DEAR MORLEY—Praise me !   I have been hard at work at Hume at Penmaenmawr, and I have got the hard part of the business—the account of his philosophy —blocked out in the bodily shape of about 180 pages foolscap MS.

But I find the job as tough as it is interesting. Hume's diamonds, before the public can see them properly, want a proper setting in a methodical and consistent shape—and that implies writing a small psychological treatise of one's own, and then cutting it down into as unobtrusive a form as possible.

So I am working away at my draught—from the point of view of an æsthetic jeweller.

As soon as I get it into such a condition as will need

only verbal trimming, I should like to have it set up in type. For it is a defect of mine that I can never judge properly of any composition of my own in manuscript.

Moreover (don't swear at this wish) I should very much like to send it to you in that shape for criticism.

The Life will be an easy business. I should like to get the book out of hand before Christmas, and will do so if possible. But my lectures begin on Tuesday, and I cannot promise.—Ever yours very faithfully,

T. H. HUXLEY.

Oct. 21, 1878.

MY DEAR MORLEY—I have received slips up to chap. ix. of Hume, and so far I do not think (saving your critical presence) that there will be much need of much modification or interpolation.

I have made all my citations from a 4-vol. edition of Hume, published by Black and Tait in 1826, which has long been in my possession.

Do you think I ought to quote Green and Grose's edition? It will be a great bother, and I really don't think that the understanding of Hume is improved by going back to eighteenth-century spelling.

I am at work upon the Life, which should not take long. But I wish that I had polished that off at Penmaenmawr as well. What with lecturing five days a week, and toiling at two anatomical monographs, it is hard to find time.

As soon as I have gone through all the eleven chapters about the Philosophy—I will send them to you and get you to come and dine some day—after you have looked at them—and go into it.—Ever yours very faithfully,

T. H. HUXLEY.

SCIENCE SCHOOLS, S. KENSINGTON,
Oct. 29, 1878.

MY DEAR MORLEY—Your letter has given me great pleasure. For though I have thoroughly enjoyed the work, and seemed to myself to have got at the heart of Hume's way of thinking, I could not tell how it would appear to others, still less could I pretend to judge of the literary form of what I had written.   And as I was quite prepared to accept your judgment if it had been unfavourable, so being what it is, I hug myself proportionately and begin to give myself airs as a man of letters.

I am through all the interesting part of Hume's life —that is, the struggling part of it—and David the successful and the fêted begins rather to bore me, as I am sorry to say most successful people do.   I hope to send the first chapter to press in another week.

Might it not be better, by the way, to divide the little book into two parts ?

Part I.—Life, Literary and Political work,

Part II.—Philosophy,

subdividing the latter into chapters or sections ?   Please tell me what you think.

I have not received the last chapter from the printer yet.   When I do I will finish revising, and then ask you to come and have a symposium over it.—Ever yours very faithfully,   T. H. HUXLEY.

P.S.—Macmillan has a lien on "The Hand."   I gave part of the lecture in another shape at Glasgow two years ago, and. M. had it reported for his magazine.   If he is good and patient he will get it in some shape some day !

4 MARLBOROUGH PLACE, N.W.,
Nov. 5, 1878.

MY DEAR MORLEY—"Davie's" philosophy is now all in print, and all but a few final pages of his biography.

So I think the time has come when that little critical symposium may take place.

Can you come and dine on Tuesday next (12) at 7 ? Or if any day except Wednesday 15th, next week, will

suit you better, it will do just as well for me.   There
will be nobody but my wife and daughters, so don't
dress.—Ever yours very faithfully,   T. H. HUXLEY.

P.S.—Will you be disgusted if in imitation of the
"English Men of Letters" I set agoing an "English Men
of Science."   Few people have any conception of the
part Englishmen have played in science, and I think it
would be both useful and interesting to bring the truth
home to the English mind.

I had about three thousand people to hear me on
Saturday at Manchester, and it would have done you
good to hear how they cheered at my allusion to personal
rule.   I had to stop and let them ease their souls.

Behold my *P.S.* is longer than my letter.   It's the
strong feminine element in my character oozing out.
"Desinit in piscem" though, and a mighty queer fish too.

> 4 MARLBOROUGH PLACE,
> *Jan.* 12, 1879.

DEAR LECKY—I am very much obliged for your
suggestion about the note at p. 9.   I am ashamed to say
that though the eleven day correction was familiar enough
to me, I had never thought about the shifting of the
beginning of the year till you mentioned it.   It is a law
of nature, I believe, that when a man says what he need
not say he is sure to blunder.   The note shall go out.

All I know about Sprat is as the author of a dull
history of the Royal Society, so I was surprised to meet
with Hume's estimate of him.

No doubt about the general hatred of the Scotch, but
you will observe that I make Millar responsible for the
peace-making assurance.

What you said to me in conversation some time ago
led me to look at Hume's position as a moralist with
some care, and I quoted the passage at p. 206 that no
doubt might be left on the matter.

The little book threatened to grow to an undue length, and therefore the question of morals is treated more briefly than was perhaps desirable.—Ever yours very faithfully,   T. H. HUXLEY.

Early in November I find the first reference to a proposed, but never completed, "English Men of Science" series in the letter to Mr. Morley above. The following letters, especially those to Sir H. Roscoe, with whom he was concerting the series, give some idea of its scope :—

4 MARLBOROUGH PLACE, N.W.
Dec. 10, 1878.

MY DEAR ROSCOE—You will think that I have broken out into letter-writing in a very unwonted fashion, but I forgot half of what I had to say this morning.

After a good deal of consultation with Macmillans, who were anxious that the "English Men of Science" series should not be too extensive, I have arranged the books as follows :—

1. Roger Bacon.
2. Harvey and the Physiologists of the 17th century.
3. Robert Boyle and the Royal Society.
4. Isaac Newton.
5. Charles Darwin.
6. English Physicists, Gilbert, Young, Faraday, Joule.
7. English Chemists, Black, Priestley, Cavendish, Davy, Dalton.
8. English Physiologists and Zoologists of the 18th century, Hunter, etc.
9. English Botanists, Ray, Crew, Hales, Brown.
10. English Geologists, Hutton, Smith, Lyell.

We may throw in the astronomers if the thing goes.
Green of Leeds will undertake 10 ; Dyer, with Hooker's
aid, 9 ; M. Foster 8 ; and I look to you for 7.

Tyndall has half promised to do Boyle, and I hope he
will. Clerk-Maxwell can't undertake Newton, and hints
X. But I won't have X.—he is too much of a bolter to
go into the tandem. I am thinking of asking Moulton,
who is strongly recommended by Spottiswoode, and is a
very able fellow, likely to put his strength into it.

Do you know anything about Chrystal of St.
Andrews ?[1] I forget whether I asked you before. From
all I hear of him I expect he would do No. 6 very well.
I have written to Adamson by this post.

I shall get off with Harvey and Darwin to my share.—
Ever yours very faithfully,   T. H. HUXLEY.

4 MARLBOROUGH PLACE, N.W.,
Dec. 26, 1878.

MY DEAR ROSCOE—I was very loth to lump the
chemists together, but Max was very strong about not
having too many books in the series ; and on the other
hand, I had my doubts how far the chemists were capable
of "dissociation" without making the book too technical.

But I do not regard the present arrangement as un-
alterable, and if you think the early chemists and the
later chemists would do better in two separate groups,
the matter is quite open to consideration.

Maxwell says he is overdone with work already, and
altogether declines to take anything new. I shall have
to look about me for a man to do the Physikers.

Of course Adamson will have to take in a view of the
science of the Middle Ages. That will be one of the
most interesting parts of the book, and I hope he will do
it well. I suppose he knows his Dante.

------

[1] Now Professor of Mathematics at Edinburgh.

The final cause of boys is to catch something or other. I trust that yours is demeasling himself properly.—Ever yours very faithfully,   T. H. HUXLEY

4 MARLBOROUGH PLACE,
Dec. 1878.

MY DEAR TYNDALL—I consider your saying the other evening that you would see "any one else d—d first," before you would assent to the little proposal I made to you, as the most distinct and binding acceptance you are capable of.   You have nothing else to swear by, and so you swear at everybody but me when you want to pledge yourself.

It will release me of an immense difficulty if you will undertake R. Boyle and the Royal Society (which of course includes Hooke) ; and the subject is a capital one.

The book should not exceed about 200 pages, and you need not be ready before this time next year.   There could not be a more refreshing piece of work just to enliven the *dolce far niente* of the Bel Alp.   (That is quite *à la* Knowles, and I begin to think I have some faculty as an editor.)

Settle your own terms with Macmillan.   They will be as joyful as I shall be to know you are going to take part in the enterprise.—Ever yours faithfully,

T. H. HUXLEY.

4 MARLBOROUGH PLACE,
Dec. 31, 1878.

MY DEAR TYNDALL—I would sooner have your Boyle, however long we may have to wait for it, than anybody else's d—d simmer.   (Now that's a "goak," and you must ask Mrs. Tyndall to explain it to you.)

Two years will I give you from this blessed New Year's eve, 1878, and if it isn't done on New Year's Day 1881 you shall not be admitted to the company of the blessed, but your dinner shall be sent to you between two plates to the most pestiferous corner of the laboratory of the Royal Institution. I am very glad you will undertake the job, and feel that I have a proper New Year's gift.

By the way, you ought to have had Hume ere this. Macmillan sent me two or three copies, just to keep his word, on Christmas Day, and I thought I should have a lot more at once.

But there is no sign—not even an advertisement—and I don't know what has become of the edition. Perhaps the bishops have bought it up.—With all good wishes, Ever yours,   T. H. HUXLEY.

Two letters—both to Tyndall—show his solicitude for his friends. The one speaks of a last and unavailing attempt made by W. K. Clifford's friends to save his life by sending him on a voyage (he died not long after at Madeira); the other urges Tyndall himself to be careful of his health.

4 MARLBOROUGH PLACE,
April 2, 1878.

MY DEAR TYNDALL—We had a sort of council about Clifford at Clark's house yesterday morning—H. Thompson, Corfield, Payne, Pollock, and myself, and I am sure you will be glad to hear the result.

From the full statement of the nature of his case made by Clark and Corfield, it appears that though grave enough in all conscience, it is not so bad as it might be, and that there is a chance, I might almost say a fair chance, for him yet. It appears that the lung mischief has never gone so far as the formation of a cavity, and that it is at present quiescent, and no other organic disease is discoverable. The alarming symptom is a general prostration—very sadly obvious when he was with us on Sunday—which, as I understand, rather renders him specially obnoxious to a sudden and rapid development of the lung disease than is itself to be feared.

It was agreed that they should go at once to Gibraltar by the P. and O., and report progress when he gets there. If strong enough he is to go on a cruise round the Mediterranean, and if he improves by this he is to go away for a year to Bogota (in S. America), which appears to be a favourable climate for such cases as his.

If he gets worse he can but return. I have done my best to impress upon him and his wife the necessity of extreme care, and I hope they will be wise.

It is very pleasant to find how good and cordial everybody is, helpful in word and deed to the poor young people. I know it will rejoice the cockles of your generous old heart to hear it.

As for yourself, I trust you are mending and allowing yourself to be taken care of by your household goddess.

With our united love to her and yourself,—Ever yours faithfully,   T. H. Huxley.

I sent your cheque to Yeo.

May, 1878.

My dear Tyndall—You were very much wanted on Saturday, as your wife will have told you, but for all that I would not have had you come on any account. You want a thorough long rest and freedom from excitement of all sorts, and I am rejoiced to hear that you are going out of the hurly-burly of London as soon as possible; and, not to be uncivil, I do hope you will stay away as long as possible, and not be deluded into taking up anything exciting as soon as you feel lively again among your mountains.

Pray give up Dublin. If you don't, I declare I will try if I have enough influence with the council to get you turned out of your office of Lecturer, and superseded.

Do seriously consider this, as you will be undoing the good results of your summer's rest. I believe your heart is as sound as your watch was when you went on your memorable slide,[1] but if you go slithering down avalanches of work and worry you can't always expect to pick up "the little creature" none the worse. The apparatus is by one of the best makers, but it has been some years in use, and can't be expected to stand rough work.

You will be glad to hear that we had cheerier news of Clifford on Saturday. He was distinctly better, and setting out on his Mediterranean voyage.—Ever yours very faithfully, T. H. HUXLEY.

A birthday letter to his son concludes the year :—

4 MARLBOROUGH PLACE, N.W.,
Dec. 10, 1878.

Your mother reminds me that to-morrow is your eighteenth birthday, and though I know that my " happy returns" will reach you a few hours too late, I cannot but send them.

You are touching manhood now, my dear laddie, and I trust that as a man your mother and I may always find reason to regard you as we have done throughout your boyhood.

The great thing in the world is not so much to seek happiness as to earn peace and self-respect. I have not troubled you much with paternal didactics—but that bit is "ower true" and worth thinking over.

---

[1] On the Piz Morteratsch ; *Hours of Exercise in the Alps*, by J. Tyndall, ch. xix.

CHAPTER XI

1879

MUCH of the work noted down for 1878 reappears
in my father's list for 1879. He was still at work
upon, or meditating his Crayfish, his Introduction to
Psychology, the Spirula Memoir, and a new edition
of the Elementary Physiology. Professor H. N.
Martin writes about the changes necessary for adapting the "Practical Biology" to American needs; the
article on Harvey was waiting to be put into
permanent form. Besides giving an address at the
Working Men's College, he lectured on Sensation and
the Uniformity of the Sensiferous Organs (*Coll. Ess.*
vi.), at the Royal Institution, Friday evening, March
7; and on Snakes, both at the Zoological Gardens,
June 5, and at the London Institution, Dec. 1. On
February 3 he read a paper at the Royal Society on
"The Characters of the Pelvis in the Mammalia, and
the Conclusions respecting the Origin of Mammals
which may be based on them"; and published in
Nature for November 6 a paper on "Certain Errors

Respecting the Structure of the Heart, attributed to
Aristotle."

Great interest attaches to this paper. He had
always wondered how Aristotle, in dissecting a heart,
had come to assert that it contained only three
chambers; and the desire to see for himself what
stood in the original, uncommented on by translators
who were not themselves anatomists, was one of the
chief reasons (I think the wish to read the Greek
Testament in the original was another) which
operated in making him take up the study of Greek
late in middle life. His practice was to read in his
book until he had come to ten new words; these he
looked out, parsed, and wrote down together with
their chief derivatives. This was his daily portion.

When at last he grappled with the passage in
question, he found that Aristotle had correctly
described what he saw under the special conditions
of his dissection, when the right auricle actually
appears as he described it, an enlargement of the
"great vein." So that this, at least, ought to be
removed from the list of Aristotle's errors. The same
is shown to be the case with his statements about
respiration. His own estimate of Aristotle as a
physiologist is between the panegyric of Cuvier and
the depreciation of Lewes: "he carried science a step
beyond the point at which he found it; a meritorious,
but not a miraculous, achievement." And it will
interest scholars to know that from his own experience
as a lecturer, Huxley was inclined to favour the

theory that the original manuscripts of the *Historia Animalium*, with their mingled accuracy and absurdity, were notes taken by some of his students. This essay was reprinted in *Science and Culture*, p. 180.

This year he brought out his second volume of essays on various subjects, written from 1870 to 1878, under the title of *Critiques and Addresses*, and later in the year, his long-delayed and now entirely recast *Introductory Primer* in the Science Primer Series.

6 BARNEPARK TERRACE, TEIGNMOUTH,
Sept. 12, 1879.

MY DEAR ROSCOE—I send you by this post my long-promised Primer, and a like set of sheets goes to Stewart.[1]

You will see that it is quite different from my first sketch, Geikie's primer having cut me out of that line—but *I* think it much better.

You will see that the idea is to develop Science out of common observation, and to lead up to Physics, Chemistry, Biology, and Psychology.

I want the thing to be good as far as it goes, so don't spare criticism.—Ever yours very faithfully,

T. H. HUXLEY

Best remembrances from us all, which we are jolly.

To his other duties he now added that of a Governor of Eton College, a post which he held till 1888, when, after doing what he could to advance progressive ideas of education, and in particular, getting a scheme adopted for making drawing part

---

[1] Balfour Stewart, Professor of Natural Philosophy in Owens College, Manchester.

of the regular curriculum, ill-health compelled him
to resign.

As for other pressure of work (he writes to Dr. Dohrn,
February 16), with the exception of the Zoological Society,
I never have anything to do with the affairs of any society
but the Royal now—I find the latter takes up all my
disposable time.   Take comfort from me.   I find 53
to be a very youthful period of existence.   I have been
better physically, and worked harder mentally, this last
twelvemonth than in any year of my life.   So a mere
boy, not yet 40 like you, may look to the future
hopefully.

From about this time dates the inception of a
short-lived society, to be called the Association of
Liberal Thinkers.   It had first taken shape in the
course of a conversation at Prof. W. K. Clifford's
house; the chief promoter and organiser being a well-
known Theistic preacher, while on the council were
men of science, critics, and scholars in various branches
of learning.   Huxley was chosen President, and the
first meeting of officers and council took place at his
house on January 25.

Professor G. J. Romanes was asked to join, but
refused on the ground that even if the negations
which he supposed the society would promulgate,
were true, it was not expedient to offer them to the
multitude.   To this Huxley wrote the following
reply (January 2, 1879) :—

Many thanks for your letter.   I think it is desirable
to explain that our Society is by no means intended to

constitute a propaganda of negations, but rather to serve as a centre of free thought.

Of course I have not a word to say in respect of your decision. I quite appreciate your view of the matter, though it is diametrically opposed to my own conviction that the more rapidly truth is spread among mankind the better it will be for them.

Only let us be sure that it is truth.

However, a course of action was proposed which by no means commended itself to several members of the council. Tyndall begs Huxley "not to commit us to a venture of the kind unless you see clearly that it meets a public need, and that it will be worked by able men," and on February 6 the latter writes—

After careful consideration of the whole circumstances of the case, I have definitely arrived at the conclusion that it is not expedient to go on with the undertaking.

I therefore resign my Presidency, and I will ask you to be so good as to intimate my withdrawal from the association to my colleagues.

In spite of having long ago "burned his ships" with regard to both the great Universities, Huxley was agreeably surprised by a new sign of the times from Cambridge. The University now followed up its recognition of Darwin two years before, by offering Huxley an honorary degree, an event of which he wrote to Professor Baynes on June 9 :—

I shall be glorious in a red gown at Cambridge to-morrow, and hereafter look to be treated as a PERSON OF RESPECTABILITY.

I have done my best to avoid that misfortune, but it's of no use.

A curious coincidence occurred here. Mr. Sandys, the public orator,[1] in his speech presenting him for the degree, picked out one of his characteristics for description in the Horatian phrase, "Propositi tenax." Now this was the family motto; and Huxley wrote to point out the coincidence :—

SCIENCE AND ART DEPARTMENT,
SOUTH KENSINGTON, *June* 11, 1879.

MY DEAR MR. SANDYS—I beg your acceptance of the inclosed photograph, which is certainly the best ever executed of me.

And by way of a memento of the claim which you

---

[1] The speech delivered by the public orator on this occasion (June 10, 1879) ran as follows :—Academi inter silvas qui verum quaerunt, non modo ipsi veritatis lumine vitam hanc umbratilem illustrare conantur, sed illustrissimum quemque veritatis investigatorem aliunde delatum ea qua par est comitate excipiunt. Adest vir cui in veritate exploranda ampla sane provincia contigit, qui sive in animantium sive in arborum et herbarum genere quicquid vivit investigat, ipsum illud vivere quid sit, quali ex origine natum sit ; qui exquirit quae cognationis necessitudo inter priores illas viventium species et has quae etiam nunc supersunt, intercedat. Olim in Oceano Australi, ubi rectis "oculis monstra natantia" vidit, victoriam prope primam, velut alter Perseus, a Medusa reportavit ; varias deinceps animantium formas quasi ab ipsa Gorgone in saxum versas sagacitate singulari explicavit ; vitae denique universae explorandae vitam suam totam dedicavit. Physicorum inter principes diu honoratus, idem (ut verbum mutuemur a Cartesio illo cujus laudes ipse in hac urbe quondam praedicavit) etiam "metaphysica" honore debito prosecutus est. Illum demum liberaliter educatum esse existimat qui cum ceteris animi et corporis dotibus instructus sit, tum praesertim quicquid turpe sit oderit, quicquid sive in arte sive in rerum natura pulchrum sit diligat ; neque tamen ipse (ut ait Aristoteles) "animalium parum pulchrorum contemplationem fastidio puerili reformidat" ; sed in perpetua animantium serie hominis vestigia perscrutari conatus, satis ampla liberalitate in universa rerum natura "humani nihil a se alienum putat." Duco ad vos virum intrepidum, facundum, propositi tenacem, Thomam Henricum Huxley.

established not only to the eloquence but also the insight of a prophet, I have added an impression of the seal with "Tenax propositi" writ plain, if not large. As I mentioned to you, it belonged to my eldest brother, who has been dead for many years. I trust that the Heralds' College may be as well satisfied as he was about his right to the coat of arms and crest.

My own genealogical inquiries have taken me so far back that I confess the later stages do not interest me.— Ever yours very faithfully,   T. H. HUXLEY.

The British Association met at Sheffield in 1879, and Huxley took this occasion to "eat the leek" in the matter of Bathybius (see vol. i. p. 427). It must be remembered that his original interpretation of the phenomenon did not involve any new theory of the origin of life, and was not put forward because of its supposed harmony with Darwin's speculations.[1]

In supporting a vote of thanks to Dr. Allman, the President, for his address, he said (see *Nature*, Aug. 28, 1879):—

I will ask you to allow me to say one word rather upon my own account, in order to prevent a misconception which, I think, might arise, and which I should regret if it did arise. I daresay that no one in this room, who

---

[1] "That which interested me in the matter was the apparent analogy of *Bathybius* with other well-known forms of lower life, such as the plasmodia of the Myxomycetes and the Rhizopods. Speculative hopes or fears had nothing to do with the matter ; and if *Bathybius* were brought up alive from the bottom of the Atlantic to-morrow, the fact would not have the slightest bearing, that I can discern, upon Mr. Darwin's speculations, or upon any of the disputed problems of biology. It would merely be one elementary organism the more added to the thousands already known." (*Coll. Ess.* v. 154.)

has attained middle life, has been so fortunate as to reach that age without being obliged, now and then, to look back upon some acquaintance, or, it may be, intimate ally of his youth, who has not quite verified the promises of that youth. Nay, let us suppose he has done quite the reverse, and has become a very questionable sort of character, and a person whose acquaintance does not seem quite so desirable as it was in those young days; his way and yours have separated; you have not heard much about him; but eminently trustworthy persons have assured you he has done this, that, or the other; and is more or less of a black sheep, in fact. The President, in an early part of his address, alluded to a certain thing— I hardly know whether I ought to call it a thing or not —of which he gave you the name Bathybius, and he stated, with perfect justice, that I had brought that thing into notice; at any rate, indeed, I christened it, and I am, in a certain sense, its earliest friend. For some time after that interesting Bathybius was launched into the world, a number of admirable persons took the little thing by the hand, and made very much of it, and as the President was good enough to tell you, I am glad to be able to repeat and verify all the statements, as a matter' of fact, which I had ventured to make about it. And so things went on, and I thought my young friend Bathybius would turn out a credit to me. But I am sorry to say, as time has gone on, he has not altogether verified the promise of his youth.

In the first place, as the President told you, he could not be found when he was wanted; and in the second place, when ho was found, all sorts of things were said about him. Indeed, I regret to be obliged to tell you that some persons of severe minds went so far as to say that he was nothing but simply a gelatinous precipitate of slime, which had carried down organic matter. If that is so, I am very sorry for it, for whoever may have joined in this error, I am undoubtedly primarily responsible for it. But I do not know at the present time of my own knowledge how the matter stands. Nothing would please me more than to investigate the matter afresh in the way it ought to be investigated, but that would require a voyage of some time, and the investigation of this thing in its native haunts is a kind of work for which, for many years past, I have had no opportunity, and which I do not think I am very likely to enjoy again. Therefore my own judgment is in an absolute state of suspension about it. I can only assure you what has been said about this friend of mine, but I cannot say whether what is said is justified or not. But I feel very happy about the matter. There is one thing about us men of science, and that is, no one who has the greatest prejudice against science can venture to say that we ever endeavour to conceal each other's mistakes. And, therefore, I rest in the most entire and complete confidence that if this should happen to be a blunder of mine, some day or other it will be carefully exposed by somebody. But pray let me remind you whether all this story about Bathybius be right or wrong, makes not the slightest difference to the general argument of the remarkable address put before you to-night. All the statements your President has made are just as true, as profoundly true, as if this little eccentric Bathybius did not exist at all.

Several letters of miscellaneous interest may be quoted.

The following acknowledges the receipt of *Essays in Romance :—*

4 MARLBOROUGH PLACE, LONDON, N.W.,
January 1879.

MY DEAR SKELTON—Being the most procrastinating letter-writer in existence, I thought, or pretended to think, when I received your *Essays in Romance* that it would not be decent to thank you until I had read the

book. And when I had done myself that pleasure, I
further pretended to think that it would be much better
to wait till I could send you my Hume book, which, as
it contains a biography, is the nearest approach to a work
of fiction of which I have yet been guilty.

The "Hume" was sent, and I hope reached you a
week ago, and as my conscience just now inquired in a
very sneering and unpleasant tone whether I had any
further pretence for not writing on hand, I thought I
might as well stop her mouth at once.

You will see oddly enough that I have answered your
question about dreams in a sort of way on page 96.[1]

You will get nothing but praise for your book, and I
shall be vilipended for mine. Is that fact, or is it not,
an evidence of a special Providence and Divine Government ?

Pray remember me very kindly to Mrs. Skelton. I
hope your interrupted visit will yet become a fact. We
have a clean bill of health now.—Ever yours very faithfully,　　　　　　　　　　　　　　T. H. HUXLEY.

SCOTTISH UNIVERSITY COMMISSION,
31 QUEEN STREET, EDINBURGH, *April* 2, 1879.

MY DEAR SKELTON—I shall be delighted to dine with
you on Wednesday, and take part in any discussion either
moral or immoral that may be started.—Ever yours very
faithfully,　　　　　　　　　　　　T. H. HUXLEY.

March 15, 1879.

MY DEAR MRS. TYNDALL—Your hearty letter is as
good as a bottle of the best sunshine. Yes, I will lunch
with you on Friday with pleasure, and Jess proposes to
attend on the occasion. . . . Her husband is in Gloucester,
and so doesn't count. The absurd creature declares she
must go back to him on Saturday—stuff and sentiment.

---

[1] Cp. *Essays in Romance*, p. 329 ; Huxley's *Hume*, p. 96.

She has only been here six or seven weeks. There is nothing said in Scripture about a wife cleaving to her husband !—With all our loves, ever yours very sincerely,

T. H. HUXLEY.

The next is to his son, then at St. Andrews University, on winning a scholarship tenable at Oxford.

SOUTH KENSINGTON, *April* 21, 1879.

MY DEAR BOY—I was very glad to get your good news this morning, and I need not tell you whether M—— was pleased or not.

But the light of nature doth not inform us of the value and duration of the "Guthrie"—and from a low and material point of view I should like to be informed on that subject. However, this is "mere matter of detail" as the Irishman said when he was asked *how* he had killed his landlord. The pleasure to us is that you have made good use of your opportunities, and finished this first stage of your journey so creditably.

I am about to write to the Master of Balliol for advice as to your future proceedings. In the meanwhile, go in for the enjoyment of your holiday with a light heart. You have earned it.—Ever your loving father,

T. H. HUXLEY.

The following, to Mrs. Clifford, was called forth by a hitch in respect to the grant to her of a Civil List pension after the death of her husband :—

4 MARLBOROUGH PLACE,
July 19, 1879.

MY DEAR LUCY—I am just off to Gloucester to fetch M—— back, and I shall have a long talk with that sage little woman over your letter.

In the meanwhile keep quiet and do nothing. I feel the force of what you say very strongly—so strongly, in fact, that I must morally ice myself and get my judgment clear and cool before I advise you what is to be done.

I am very sorry to hear you have been so ill. For the present dismiss the matter from your thoughts and give your mind to getting better. Leave it all to be turned over in the mind of that cold-blooded, worldly, cynical old fellow, who signs himself—Your affectionate

PATER.

The last is to Mr. Edward Clodd, on receiving his book *Jesus of Nazareth.*

4 MARLBOROUGH PLACE, ABBEY ROAD, N.W.,
Dec. 21, 1879.

MY DEAR MR. CLODD—I have been spending all this Sunday afternoon over the book you have been kind enough to send me, and being a swift reader, I have travelled honestly from cover to cover.

It is the book I have been longing to see ; in spirit, matter and form it appears to me to be exactly what people like myself have been wanting. For though for the last quarter of a century I have done all that lay in my power to oppose and destroy the idolatrous accretions of Judaism and Christianity, I have never had the slightest sympathy with those who, as the Germans say, would "throw the child away along with the bath"— and when I was a member of the London School Board I fought for the retention of the Bible, to the great scandal of some of my Liberal friends—who can't make out to this day whether I was a hypocrite, or simply a fool on that occasion.

But my meaning was that the mass of the people should not be deprived of the one great literature which is open to them—not shut out from the perception of their relations with the whole past history of civilised mankind—not excluded from such a view of Judaism and Jesus of Nazareth as that which at last you have given us.

I cannot doubt that your work will have a great success not only in the grosser, but the better sense of the word.—I am yours very faithfully,   T. H. HUXLEY.

The winter of 1879-80 was memorable for its prolonged spell of cold weather. One result of this may be traced in a New Year's letter from Huxley to his eldest daughter. "I have had a capital holiday—mostly in bed—but I don't feel so grateful for it as I might do." To be forced to avoid the many interruptions and distractions of his life in London, which claimed the greater part of his time, he would regard as an unmixed blessing; as he once said feelingly to Professor Marsh, "If I could only break my leg, what a lot of scientific work I could do!" But he was less grateful for having entire inaction forced upon him.

However, he was soon about again, and wrote as follows in answer to a letter from Sir Thomas (afterwards Lord) Farrer, which called his attention, as an old Fishery Commissioner, to a recent report on the sea-fisheries.

4 MARLBOROUGH PLACE,
Jan. 9, 1880.

MY DEAR FARRER—I shall be delighted to take a dive into the unfathomable depths of official folly; but your promised document has not reached me.

Your astonishment at the tenacity of life of fallacies, permit me to say, is shockingly unphysiological. They,

like other low organisms, are independent of brains, and
only wriggle the more, the more they are smitten on the
place where the brains ought to be—I don't know B.,
but I am convinced that A. has nothing but a spinal
cord, devoid of any cerebral development. Would Mr.
Cross give him up for purposes of experiment? Lingen
and you might perhaps be got to join in a memorial to
that effect.—Ever yours very faithfully,

T. H. HUXLEY.

A fresh chapter of research, the results of which
he now began to give to the public, was the history
of the Dog. On April 6 and 13 he lectured at the
Royal Institution "On Dogs and the Problems connected with them"—their relation to other animals,
and the problem of the origin of the domestic dog,
and the dog-like animals in general. As so often
before, these lectures were the outcome of the careful
preparation of a course of instruction for his students.
The dog had been selected as one of the types of
mammalian structure upon which laboratory work
was to be done. Huxley's own dissections had led
him on to a complete survey of the genus, both wild
and domestic. As he writes to Darwin on May 10:—

I wish it were not such a long story that I could tell
you all about the dogs. They will make out such a case
for "Darwinismus" as never was. From the South
American dogs at the bottom (*C. vetulus, cancrivorus*, etc.)
to the wolves at the top, there is a regular gradual pro-
gression, the range of variation of each "species" over-
lapping the ranges of those below and above. Moreover,
as to the domestic dogs, I think I can prove that the
small dogs are modified jackals, and the big dogs ditto

wolves. I have been getting capital material from India, and working the whole affair out on the basis of measurements of skulls and teeth.

However, my paper for the Zoological Society is finished, and I hope soon to send you a copy of it. . . .

Unfortunately he never found time to complete his work for final publication in book form, and the rough, unfinished notes are all that remain of his work, beyond two monographs "On the Epipubis in the Dog and Fox" (*Proc. Roy. Soc.* xxx. 162-63), and "On the Cranial and Dental Characters of the Canidae" (*Proc. Zool. Soc.* 1880, pp. 238-288).

The following letters deal with the collection of specimens for examination :—

4 MARLBOROUGH PLACE,
Jan. 17, 1880.

MY DEAR FLOWER—I happened to get hold of two foxes this week—a fine dog fox and his vixen wife; and among other things, I have been looking up Cowper's glands, the supposed absence of which in the dogs has always "gone agin' me." Moreover, I have found them (or their representatives) in the shape of two small sacs, which open by conspicuous apertures into the urethra immediately behind the bulb. If your *Icticyon* was a male, I commend this point to your notice.

Item—If you have not already begun to macerate him, do look for the "marsupial" fibro-cartilages, which I have mentioned in my "Manual," but the existence of which blasphemers have denied. I found them again at once in both Mr. and Mrs. Vulpes. You spot them immediately by the *pectineus* which is attached to them.

The dog-fox's cæcum is so different from the vixen's that Gray would have made distinct genera of them.—Ever yours very faithfully,   T. H. HUXLEY.

4 MARLBOROUGH PLACE, N.W.,
May 2, 1880.

MY DEAR FAYRER—I am greatly obliged for the
skulls, and I hope you will offer my best thanks to your
son for the trouble he has taken in getting them.

The "fox" is especially interesting because it is not a
fox, by any manner of means, but a big jackal with some
interesting points of approximation towards the cuons.

I do not see any locality given along with the specimens. Can you supply it?

I have got together some very curious evidence of the
wider range of variability of the Indian jackal, and the
"fox" which your son has sent is the most extreme form
in one direction I have met with.

I wish I could get some examples from the Bombay
and Madras Presidencies and from Ceylon, as well as from
Central India. Almost all I have seen yet are from
Bengal.—Ever yours very faithfully,

T. H. HUXLEY.

Between the two lectures on the Dog, mentioned
above, on April 9, Huxley delivered a Friday evening
discourse, at the same place, "On the Coming of Age
of the Origin of Species" (*Col. Ess.* ii. 227). Re-
viewing the history of the theory of evolution in the
twenty-one years that had elapsed since the *Origin of
Species* first saw the light in 1859, he did not merely
dwell on the immense influence the "Origin" had
exercised upon every field of biological inquiry.
"Mere insanities and inanities have before now swollen
to portentous size in the course of twenty years."
"History warns us that it is the customary fate of
new truths to begin as heresies, and to end as super-

stitions." There was actual danger lest a new generation should "accept the main doctrines of the *Origin
of Species* with as little reflection, and it may be with
as little justification, as so many of our contemporaries, years ago, rejected them."

So dire a consummation, he declared, must be
prevented by unflinching criticism, the essence of
the scientific spirit, "for the scientific spirit is of
more value than its products, and irrationally held
truths may be more harmful than reasoned errors."

What, then, were the facts which justified so
great a change as had taken place, which had removed some of the most important qualifications
under which he himself had accepted the theory?
He proceeded to enumerate the "crushing accumulation of evidence" during this period, which had
proved the imperfection of the geological record;
had filled up enormous gaps, such as those between
birds and reptiles, vertebrates and invertebrates,
flowering and flowerless plants, or the lowest forms
of animal and plant life. More: paleontology alone
has effected so much—the fact that evolution has
taken place is so irresistibly forced upon the mind
by the study of the Tertiary mammalia brought to
light since 1859, that "if the doctrine of evolution
had not existed, paleontologists must have invented
it." He further developed the subject by reading
before the Zoological Society a paper "On the Application of the Laws of Evolution to the Arrangement
of the Vertebrata, and more particularly of the

Mammalia" (*Proc. Z. S.* 1880, pp. 649-662).   In reply to Darwin's letter thanking him for the "Coming of Age" (*Life and Letters*, iii. 24), he wrote on May 10 :—

MY DEAR DARWIN—You are the cheeriest letter-writer I know, and always help a man to think the best of his doings.

I hope you do not imagine because I had nothing to say about "Natural Selection," that I am at all weak of faith on that article.   On the contrary, I live in hope that as palæontologists work more and more in the manner of that "second Daniel come to judgment," that wise young man M. Filhal, we shall arrive at a crushing accumulation of evidence in that direction also.   But the first thing seems to me to be to drive the fact of evolution into people's heads ; when that is once safe, the rest will come easy.

I hear that *ce cher* X. is yelping about again ; but in spite of your provocative messages (which Rachel retailed with great glee), I am not going to attack him nor anybody else.

Another popular lecture on a zoological subject was that of July 1 on "Cuttlefish and Squids," the last of the "Davis" lectures given by him at the Zoological Gardens.

More important were two other essays delivered this year.   The "Method of Zadig" (*Coll. Ess.* iv. 1), an address at the Working Men's College, takes for its text Voltaire's story of the philosopher at the Oriental court, who, by taking note of trivial indications, obtains a perilous knowledge of things which his neighbours ascribe either to thievery or magic.

This introduces a discourse on the identity of the methods of science and of the judgments of common life, a fact which, twenty-six years before, he had briefly stated in the words, "Science is nothing but trained and organised common sense" (*Coll. Ess.* iii. 45).

The other is "Science and Culture" (*Coll. Ess.* iii. 134), which was delivered on October 1, as the opening address of the Josiah Mason College at Birmingham, and gave its name to a volume of essays published in the following year. Here was a great school founded by a successful manufacturer, which was designed to give an education at once practical and liberal, such as the experience of its founder approved, to young men who meant to embark upon practical life. A "mere" literary training—*i.e.* in the classical languages—was excluded, but not so the study of English literature and modern languages. The greatest stress was laid on training in the scientific theory and practice on which depend the future of the great manufactures of the north.

The question dealt with in this address is whether such an education can give the culture demanded of an educated man to-day. The answer is emphatically Yes. English literature is a field of culture second to none, and for solely literary purposes, a thorough knowledge of it, backed by some other modern language, will amply suffice. Combined with this, a knowledge of modern science, its principles and results, which have so profoundly modified society and have created modern civilisation, will give a "criticism of life," as Matthew Arnold defined "the end and aim of all literature," that is to say culture, unattainable by any form of education which neglects it. In short, although the "culture" of former periods might be purely literary, that of to-day must be based, to a great extent, upon natural science.

This autumn several letters passed between him and Darwin. The latter, contrary to his usual custom, wrote a letter to *Nature*, in reply to an unfair attack which had been made upon evolution by Sir Wyville Thomson in his Introduction to *The Voyage of the Challenger* (see Darwin, *Life and Letters*, iii. 242), and asked Huxley to look over the concluding sentences of the letter, and to decide whether they should go with the rest to the printer or not. "My request," he writes (Nov. 5), "will not cost you much trouble—*i.e.* to read two pages—for I know that you can decide at once." Huxley struck them out, replying on the 14th, "Your pinned-on paragraph was so good that, if I had written it myself, I should have been unable to refrain from sending it on to the printer. But it is much easier to be virtuous on other people's account; and though Thomson deserved it and more, I thought it would be better to refrain. If I say a savage thing, it is only 'Pretty Fanny's way'; but if you do, it is not likely to be forgotten."

The rest of this correspondence has to do with a plan of Darwin's, generous as ever, to obtain a Civil List pension for the veteran naturalist, Wallace, whose magnificent work for science had brought him but little material return.   He wrote to consult Huxley as to what steps had best be taken; the latter replied in the letter of November 14 :—

The papers *in re* Wallace have arrived, and I lose no time in assuring you that all my "might, amity, and authority," as Essex said when that sneak Bacon asked him for a favour, shall be exercised as you wish.

On December 11 he sends Darwin the draft of a memorial on the subject, and on the 28th suggests that the best way of moving the official world would be for Darwin himself to send the memorial, with a note of his own, to Mr. Gladstone, who was then Prime Minister and First Lord of the Treasury :—

Mr. G. can do a thing gracefully when he is so minded, and unless I greatly mistake, he will be so minded if you write to him.

The result was all that could be hoped.   On January 7 Darwin writes :—"Hurrah! hurrah! read the enclosed.   Was it not extraordinarily kind of Mr. Gladstone to write himself at the present time? . . . I have written to Wallace.   He owes much to you.   Had it not been for your advice and assistance, I should never have had courage to go on."

The rest of the letter to Darwin of December 28 is characteristic of his own view of life.   As he wrote four years before (see p. 216), he was no pessimist any more than he was a professed optimist. If the vast amount of inevitable suffering precluded the one view, the gratuitous pleasures, so to speak, of life preclude the other. Life properly lived is worth living, and would be even if a malevolent fate had decreed that one should suffer, say, the pangs of toothache two hours out of every twenty-four. So he writes :—

We have had all the chicks (and the husbands of such as are therewith provided) round the Christmas table once more, and a pleasant sight they were, though I say it that shouldn't. Only the grand-daughter left out, the young woman not having reached the age when change and society are valuable.

I don't know what you think about anniversaries. I like them, being always minded to drink my cup of life to the bottom, and take my chance of the sweets and bitters.

The following is to his Edinburgh friend Dr. Skelton, whose appreciation of his frequent companionship had found outspoken expression in the pages of *The Crookit Meg*.

4 MARLBOROUGH PLACE, N.W.,
Nov. 14, 1880.

MY DEAR SKELTON—When the *Crooked Meg* reached me I made up my mind that it would be a shame to send the empty acknowledgment which I give (or don't give) for most books that reach me.

But I am over head and ears in work—time utterly wasted in mere knowledge getting and giving—and for six weeks not an hour for real edification with a wholesome story.

But this Sunday afternoon being, by the blessing of God, as beastly a November day as you shall see, I have attended to my spiritual side and been visited by a blessing in the shape of some very pretty and unexpected words anent mysel'.[1]

In truth, it is right excellent story, though, being distinctly in love with Eppie, I can only wonder how you had the heart to treat her so ill. A girl like that should have had two husbands—one "wisely ranged for show" and t'other *de par amours.*

Don't ruin me with Mrs. Skelton by repeating this, but please remember me very kindly to her.—Ever yours very faithfully,   T. H. HUXLEY.

The following letter to Tyndall was called forth by an incident in connection with the starting of the *Nineteenth Century.* Huxley had promised to help the editor by looking over the proofs of a monthly article on contemporary science. But his advertised position as merely adviser in this to the editor was overlooked by some who resented what they supposed to be his assumption of the rôle of critic in general

---

[1] The passage referred to stands on p. 72 of *The Crookit Meg*, and describes the village naturalist and philosopher, Adam Meldrum, "who in his working hours cobbled old boats, and knew by heart the plays of Shakespeare and the *Pseudodoxia Epidemica* of Sir Thomas Browne."

"For the rest it will be enough to add that this long, gaunt, bony cobbler of old boats was—was—(may I take the liberty, Mr. Professor ?) a village Huxley of the year One. The colourless brilliancy of the great teacher's style, the easy facility with which the drop of light forms itself into a perfect sphere as it falls from his pen, belong indeed to a consummate master of the art of expression, which Adam of course was not ; but the mental lucidity, justice, and balance, as well as the reserve of power, and the Shakespearian gaiety of touch, which made the old man one of the most delightful companions in the world, were essentially Huxleian."

to his fellow-workers in science. At a meeting of
the x Club, Tyndall made a jesting allusion to this;
Huxley, however, thought the mere suggestion too
grave for a joke, and replied with all seriousness to
clear himself from the possibility of such misconception. And the same evening he wrote to Tyndall :—

ATHENÆUM CLUB, PALL MALL, S.W.,
Dec. 2, 1880.

MY DEAR TYNDALL—I must tell you the ins and outs
of this *Nineteenth Century* business. I was anxious to
help Knowles when he started the journal, and at his
earnest and pressing request I agreed to do what I have
done. But being quite aware of the misinterpretation to
which I should be liable if my name " sans phrase " were
attached to the article, I insisted upon the exact words
which you will find at the head of it ; and which seemed,
and still seem to me, to define my position as a mere
adviser of the editor.

Moreover, by diligently excluding any expression of
opinion on the part of the writers of the compilation, I
thought that nobody could possibly suspect me of assuming the position of an authority even on the subjects with
which I may be supposed to be acquainted, let alone
those such as physics and chemistry, of which I know no
more than any one of the public may know.

Therefore your remarks came upon me to-night with
the sort of painful surprise which a man feels who is
accused of the particular sin of which he flatters himself
he is especially *not guilty*, and " roused my corruption "
as the Scotch' have it. But there is no need to say
anything about that, for you were generous and good as
I have always found you. Only I pray you, if hereafter
it strikes you that any doing of mine should be altered
or amended, tell me yourself and privately, and I promise

you a very patient listener, and what is more a very thankful one.—Ever yours, T. H. HUXLEY.

Tyndall replied with no less frankness, thanking him for the friendly promptitude of his letter, and explaining that he had meant to speak privately on the matter, but had been forestalled by the subject coming up when it did. And he wound up by declaring that it would be too absurd to admit the power of such an occasion "to put even a momentary strain upon the cable which has held us together for nine and twenty years."

At the very end of the year, George Eliot died. A proposal was immediately set on foot to inter her remains in Westminster Abbey, and various men of letters pressed the matter on the Dean, who was unwilling to stir without a very strong and general expression of opinion. To Mr. Herbert Spencer, who had urged him to join in memorialising the Dean, Huxley replied as follows :—

4 MARLBOROUGH PLACE,
Dec. 27, 1880.

MY DEAR SPENCER—Your telegram which reached me on Friday evening caused me great perplexity, inasmuch as I had just been talking with Morley, and agreeing with him that the proposal for a funeral in Westminster Abbey had a very questionable look to us, who desired nothing so much as that peace and honour should attend George Eliot to her grave.

It can hardly be doubted that the proposal will be bitterly opposed, possibly (as happened in Mill's case with less provocation) with the raking up of past histories, about which the opinion even of those who have least the desire or the right to be pharisaical is strongly divided, and which had better be forgotten.

With respect to putting pressure on the Dean of Westminster, I have to consider that he has some confidence in me, and before asking him to do something for which he is pretty sure to be violently assailed, I have to ask myself whether I really think it a right thing for a man in his position to do.

Now I cannot say I do. However much I may lament the circumstance, Westminster Abbey is a Christian Church and not a Pantheon, and the Dean thereof is officially a Christian priest, and we ask him to bestow exceptional Christian honours by this burial in the Abbey. George Eliot is known not only as a great writer, but as a person whose life and opinions were in notorious antagonism to Christian practice in regard to marriage, and Christian theory in regard to dogma. How am I to tell the Dean that I think he ought to read over the body of a person who did not repent of what the Church considers mortal sin, a service not one solitary proposition in which she would have accepted for truth while she was alive? How am I to urge him to do that which, if I were in his place, I should most emphatically refuse to do?

You tell me that Mrs. Cross wished for the funeral in the Abbey. While I desire to entertain the greatest respect for her wishes, I am very sorry to hear it. I do not understand the feeling which could create such a desire on any personal grounds, save those of affection, and the natural yearning to be near even in death to those whom we have loved. And on public grounds the wish is still less intelligible to me. One cannot eat one's cake and have it too. Those who elect to be free in thought and deed must not hanker after the rewards, if they are to be so called, which the world offers to those who put up with its fetters.

Thus, however I look at the proposal it seems to me to be a profound mistake, and I can have nothing to do with it.

I shall be deeply grieved if this resolution is ascribed to any other motives than those which I have set forth at more length than I intended.—Ever yours very faithfully,   T. H. HUXLEY.

CHAPTER XII

1881

THE last ten years had found Huxley gradually involved more and more in official duties. Now, with the beginning of 1881, he became yet more deeply engrossed in practical and administrative work, more completely cut off from his favourite investigations, by his appointment to an Inspectorship of Fisheries, in succession to the late Frank Buckland. It is almost pathetic to note how he snatched at any spare moments for biological research. No sooner was a long afternoon's work at the Home Office done, than, as Professor Howes relates, he would often take a hansom to the laboratory at South Kensington, and spend a last half-hour at his dissections before going home.

The Inspectorship, which was worth £700 a year, he held in addition to his post at South Kensington, the official description of which now underwent another change. In the first place, his official connection with the Survey appears to have ceased this year, the last report made by him being in 1881.

His name, however, still appeared in connection with the post of Naturalist until his retirement in 1885, and it was understood that his services continued to be available if required. Next, in October of this year, the Royal School of Mines was incorporated with the newly established Normal School—or as it was called in 1890, Royal College of Science, and the title of Lecturer on General Natural History was suppressed, and Huxley became Professor of Biology and Dean of the College at a salary of £800, for it was arranged on his appointment to the Inspectorship, that he should not receive the salary attached to the post of Dean. Thus the Treasury saved £200 a year.

As Professor of Biology, he was under the Lord President of the Council; as Inspector of Fisheries, under the Board of Trade; hence some time passed in arranging the claims of the two departments before the appointment was officially made known, as may be gathered from the following letters :—

To Sir John Donnelly

4 Marlborough Place,
Dec. 27, 1880.

My dear Donnelly—I tried hard to have a bad cold last night, and though I blocked him with quinine, I think I may as well give myself the benefit of the Bank Holiday and keep the house to-day.

There is a chance of your getting early salmon yet. I wrote to decline the post on Friday, but on Saturday evening the Home Secretary sent a note asking to see me yesterday. As he had re-opened the question, of course I felt justified in stating all the pros and cons of the case as personal to myself and my rather complicated official position. . . . He entered into the affair with a warmth and readiness which very agreeably surprised me, and he proposes making such arrangements as will not oblige me to have anything to do with the weirs or the actual inspection. Under these circumstances the post would be lovely—if I can hold it along with the other things. And of his own motion the Home Secretary is going to write to Lord Spencer about it to see if he cannot carry the whole thing through.

If this could be managed, I could get great things done in the matter of fish culture and fish diseases at South Kensington, if poor dear X.'s rattle trappery could be turned to proper account, without in any way interfering with the work of the School.

At any rate, my book stands not to lose, and may win —the innocence of the dove is not always divorced from the wisdom of the sarpent. [Sketch of the "Sarpent."]

To Lord Farrer

4 Marlborough Place,
Jan. 18, 1881.

My dear Farrer—I have waited a day or two before thanking you for your very kind letter, in the hope that I might be able to speak as one knowing where he is.

But as I am still, in an official sense, nowhere, I will not delay any longer.

I had never thought of the post, but the Home Secretary offered it to me in a very kind and considerate manner, and after some hesitation I accepted it. But some adjustment had to be made between my master, the Lord President, and the Treasury; and although everybody seems disposed to be very good to me, the business is not yet finally settled. Whence the newspapers get their information I don't know—but it is always wrong in these matters.

As you know, I have had a good apprenticeship to the work [1]—and I hope to be of some use; of the few innocent pleasures left to men past middle life—the jamming common-sense down the throats of fools is perhaps the keenest.

May we do some joint business in that way!—Ever yours very faithfully,   T. H. HUXLEY.

To his Eldest Son

Feb. 14, 1881.

I have entered upon my new duties as Fishery Inspector, but you are not to expect salmon to be much cheaper just yet.

My colleague and I have rooms at the Home Office, and I find there is more occupation than I expected, but no serious labour.

Every now and then I shall have to spend a few days in the country, holding inquiries, and as salmon rivers are all in picturesque parts of the country, I shall not object to that part of the business.

The duties of the new office were partly scientific, partly administrative. On the one hand, the natural history and diseases of fish had to be investigated; on the other, regulations had to be carried out, weirs and salmon passes approved, disputes settled, reports written. I find, for instance, that apart from the work in London, visits of inspection in all parts of

---

[1] He had already served on two Fishery Commissions, 1862 and 1864-5.

the country took up twenty-eight days between March and September this year.

Sir Spencer Walpole, who was his colleague for some years, has kindly given me an account of their work together.

Early in 1881, Sir William Harcourt appointed Professor Huxley one of Her Majesty's Inspectors of Fisheries. The office had become vacant through the untimely death, in the preceding December, of the late Mr. Frank Buckland. Under an Act, passed twenty years before, the charge of the English Salmon Fisheries had been placed under the Home Office, and the Secretary of State had been authorised to appoint two Inspectors to aid him in administering the law. The functions of the Home Office and of the Inspectors were originally simple, but they had been enlarged by an Act passed in 1873, which conferred on local conservators elaborate powers of making bye-laws for the development and preservation of the Fisheries. These bye-laws required the approval of the Secretary of State, who was necessarily dependent on the advice of his Inspectors in either allowing or disallowing them.

In addition to the nominal duties of the Inspectors, they became—by virtue of their position—the advisers of the Government on all questions connected with the Sea Fisheries of Great Britain. These fisheries are nominally under the Board of Trade, but, as this Board at that time had no machinery at its disposal for the purpose, it naturally relied on the advice of the Home Office Inspectors in all questions of difficulty, on which their experience enabled them to speak with authority.

For duties such as these, which have been thus briefly described, Professor Huxley had obvious qualifications. On all subjects relating to the Natural History of Fish he spoke with decisive authority. But, in addition to his scientific attainments, from 1863 to 1865 he had been a member of the Commission which had conducted an elaborate investigation into the condition of the Fisheries of the United Kingdom, and had taken a large share in the preparation of a Report, which—notwithstanding recent changes in law and policy—remains the ablest and most exhaustive document which has ever been laid before Parliament on the subject.

This protracted investigation had convinced Professor Huxley that the supply of fish in the deep sea was practically inexhaustible; and that, however much it might be necessary to enforce the police of the seas by protecting particular classes of sea fishermen from injury done to their instruments by the operations of other classes, the primary duty of the legislature was to develop sea fishing, and not to place restrictions on sea fishermen for any fears of an exhaustion of fish.

His scientific training, moreover, made him ridicule the modern notion that it was possible to stock the sea by artificial methods. He wrote to me, when the Fisheries Exhibition of 1883 was in contemplation, "You may have seen that we have a new Fish Culture Society. C—— talked gravely about our stocking the North Sea with cod! After that I suppose we shall take up herrings: and I mean to propose whales, which, as all the world knows, are terribly over fished!" And after the exhibition was over he wrote to me again, with reference to a report which the Commission had asked me to draw up: "I have just finished reading your report, which has given me a world of satisfaction. . . . I am particularly glad that you have put in a word of warning to the fish culturists." [1]

He was not, however, equally certain that particular

---

[1] When I was asked to write the report on this Commission, I said that I would do so if Sir E. Birkbeck, its chairman, and Professor Huxley, both met me to discuss the points to be noticed. The meeting duly took place : and I opened it by asking what was

areas of Sea Shore might not be exhausted by our fishing. He extended in 1883 an order which Mr. Buckland and I had made in 1879 for restricting the taking of crabs and lobsters on the coast of Norfolk, and he wrote to me on that occasion : " I was at Cromer and Sheringham last week, holding an inquiry for the Board of Trade about the working of your order of 1879. According to all accounts, the crabs have multiplied threefold in 1881 and 1882. Whether this is *post hoc* or *propter hoc* is more than I should like to say. But at any rate, this is a very good *primâ facie* case for continuing the order, and I shall report accordingly. Anyhow, the conditions are very favourable for a long-continued experiment in the effects of regulation, and, ten years hence, there will be some means of judging of the value of these restrictions."

If, however, Professor Huxley was strongly opposed to unnecessary interference with the labours of sea fishermen, he was well aware of the necessity of protecting migratory fish like salmon, against over-fishing : and his reports for 1882 and 1883—in which he gave elaborate accounts of the results of legislation on the Tyne and on the Severn — show that he keenly appreciated the necessity of regulating the Salmon Fisheries.

It so happened that at the time of his appointment, many of our important rivers were visited by " Saprolegnia ferax," the fungoid growth which became popularly known as Salmon Disease. Professor Huxley gave much time to the study of the conditions under which the fungus flourished : he devoted much space in his earlier reports to the subject : and he read a paper upon it at a remarkable meeting of the Royal Society in the summer of 1881. He took a keen interest in these investigations, and he wrote to me from North Wales, at the end of

the chief lesson to be drawn from the exhibition ? " Well," said Professor Huxley, " the chief lesson to be drawn from the exhibition is that London is in want of some open air amusement on summer evenings."

1881, " The salmon brought to me here have not been so badly diseased as I could have wished, and the fungus dies so rapidly out of the water that only one specimen furnished me with materials in lively condition. These I have cultivated : and to my great satisfaction have got some flies infected.   With nine precious muscoid corpses, more or less ornamented with a lovely fur trimming of Saprolegnia, I shall return to London to-morrow, and shall be ready in a short time, I hope, to furnish Salmon Disease wholesale, retail, or for exportation."

In carrying out the duties of our office, Professor Huxley and I were necessarily thrown into very close communication.   There were few days in which we did not pass some time in each other's company : there were many weeks in which we travelled together through the river basins of this country.   I think that I am justified in saying that official intercourse ripened into warm personal friendship, and that, for the many months in which we served together, we lived on terms of intimacy which are rare among colleagues or even among friends.

It is needless to say that, as a companion, Professor Huxley was the most delightful of men.   Those who have met him in society, or enjoyed the hospitality of his house, must have been conscious of the singular charm of a conversation, which was founded on knowledge, enlarged by memory, and brightened by humour.   But, admirable as he was in society, no one could have realised the full charm of his company who had not conversed with him alone.   He had the rare art of placing men, whose knowledge and intellect were inferior to his own, at their ease.   He knew how to draw out all that was best in the companion who suited him ; and he had equal pleasure in giving and receiving.   Our conversation ranged over every subject.   We discusssed together the grave problems of man and his destiny ; we disputed on the minor complications of modern politics ; we criticised one another's literary judgments ; and we laughed over

the stories which we told one another, and of which
Professor Huxley had an inexhaustible fund.

In conversation Professor Huxley displayed the quality
which distinguished him both as a writer and a public
speaker. He invariably used the right words in the right
sense. Those who are jointly responsible—as he and I
were often jointly responsible—for some written document,
have exceptional opportunities of observing this quality.
Professor Huxley could always put his finger on a wrong
word, and he always instinctively chose the right one.
It was this qualification—a much rarer one than people
imagine—which made Professor Huxley's essays clear to
the meanest understanding, and which made him, in my
judgment, the greatest master of prose of his time. The
same quality was equally observable in his spoken speech.
I happened to be present at the anniversary dinner of the
Royal Society, at which Professor Huxley made his last
speech. And, as he gave an admirable account of the
share which he had taken in defending Mr. Darwin
against his critics, I overheard the present Prime Minister[1]
say, " What a beautiful speaker he is."

In 1882, the duties of another appointment forced
me to resign the Inspectorship, which I had held for so
long : and thenceforward my residence in the Isle of Man
gave me fewer opportunities of seeing Professor Huxley :
our friendship, however, remained unbroken ; and
occasional visits to London gave me many opportunities
of renewing it. He retained his own appointment as
Inspector for more than three years after my resignation.
He served, during the closing months of his officialship,
on a Royal Commission on trawling, over which the late
Lord Dalhousie presided. But his health broke down
before the commissioners issued their report, and he was
ordered abroad. It so happened that in the spring of
1885 I was staying at Florence, when Professor and

---

[1] Lord Salisbury.

Mrs. Huxley passed through it on their way home. He
had at that time seen none of his old friends, and was
only slowly regaining strength. After his severe illness
Mrs. Huxley encouraged me to take him out for many
short walks, and I did my best to cheer him in his
depressed condition. He did not then think that he
had ten years of—on the whole—happy life before him.
He told me that he was about to retire from all his work,
and he added, that he had never enjoyed the Inspectorship after I had left it. I am happy in believing that
the remark was due to the depression from which he was
suffering, for he had written to me two years ago, "The
office would be quite perfect, if they did not want an
annual report. I can't go in for a disquisition on river
basins after the manner of Buckland, and you have
exhausted the other topics. I polished off the Salmon
Disease pretty fully last year, so what the deuce am I
to write about?"

I saw Professor Huxley for the last time on the
Christmas day before his death. I spent some hours with
him, with no other companions than Mrs. Huxley and
my daughter. I had never seen him brighter or happier,
and his rich, playful and sympathetic talk vividly recalled
the many brilliant hours which I had passed in his
company some twelve or thirteen years before.

One word more. No one could have known Professor
Huxley intimately without recognising that he delighted
in combat. He was never happier than when he was
engaged in argument or controversy, and he loved to
select antagonists worthy of his steel. The first public
inquiry which we held together was attended by a great
nobleman, whom Professor Huxley did not know by sight,
but who rose at the commencement of our proceedings to
offer some suggestions. Professor Huxley directed him
to sit down, and not interrupt the business. I told my
colleague in a whisper whom he was interrupting. And
I was amused, as we walked away to luncheon together,

by his quaint remark to me, " We have begun very well, we have sat upon a duke." [1]

If, however, a love of argument and controversy occasionally led him into hot water, I do not think that his polemical tendencies ever cost him a friend. His antagonists must have recognised the fairness of his methods, and must have been susceptible to the charm of the man. The high example which he set in controversy, moreover, was equally visible in his ordinary life. Of all the men I have ever known, his ideas and his standard were—on the whole—the highest. He recognised that the fact of his religious views imposed on him the duty of living the most upright of lives, and I am very much of the opinion of a little child, now grown into an accomplished woman, who, when she was told that Professor Huxley had no hope of future rewards, and no fear of future punishments, emphatically declared : " Then I think Professor Huxley is the best man I have ever known."

Extracts from his letters home give some further idea of the kind of work entailed. Thus in March and again in May he was in Wales, and writes :—

CROMFFYRATELLIONPTRROCH,
May 24.

Mr. Barrington's very pretty place about five miles from Abergavenny, wherein I write, may or may not have the name which I have written on at the top of the

---

[1] Of this he wrote home on March 15, 1881 : "Somebody produced the *Punch* yesterday and showed it to me, to the great satisfaction of the Duke of ——, who has attended our two meetings. I nearly had a shindy with him at starting, but sweetness and light (in my person) carried the day." This *Punch* contained the cartoon of Huxley in nautical costume riding on a salmon ; contrary to the custom of *Punch*, it made an unfair hit in appending to his name the letters £ s. d. Never was any one who deserved the imputation less.

page, as it is Welsh ; however it is probably that or something like it.   I forgot to inquire.

We are having the loveliest weather, and yesterday
went looking up weirs with more or less absurd passes up
a charming valley not far hence.   It is just seven o'clock,
and we are going to breakfast and start at eight to fit in
with the tides of the Severn.   It is not exactly clear
where we shall be to-night. . . . Now I must go to breakfast, for I got up at six.   *Figurez vous ça.*

May 29—*Hereford.*—We are favoured by the weather
again, though it is bitter cold under the bright sunshine.
We stopped at Worcester yesterday, and I went to examine
some weirs hard by.   This involved three or four miles'
country walking, and was all to the good.   If the Inspector
business were all of this sort it would be all that fancy
painted it.   We shall have a long sitting to-day. . . . [He
fears to be detained into the night by " over-fluent
witnesses."]

In April he spent several days at Norwich, in
connection with the National Fishery Exhibition held
there.

April 19.—We had a gala day yesterday. . . . The
exhibition of all manner of fish and fishing apparatus was
ready, for a wonder, and looked very well.   The Prince
and Princess arrived, and we had the usual address and
reply and march through.   Afterwards a mighty *déjeûner*
in the St. Andrew's Hall—a fine old place looking its best.
I was just opposite the Princess, and I could not help
looking at her with wonderment.   She looked so fresh
and girlish.   She came and talked to me afterwards in a
very pleasant simple way.

Walpole and I went in with our host yesterday afternoon and started to return on the understanding that he
should pick us up a few miles out.   Of course we took
the wrong road, and walked all the way, some eight miles

or so.   However, it did us good, and after a champagne
lunch we thought we could not do better than repeat the
operation yesterday.

I feel quite set up by finding that after standing about
for hours I can walk eight miles without any particular
fatigue.   Life in the old dog yet!   Walpole is a capital
companion—knows a great many things, and talks well
about them, so we get over the ground pleasantly.

April 20.—There was a long day of it yesterday looking over things in the Exhibition till late in the afternoon,
and then a mighty dinner in St. Andrew's Hall given
by a Piscatorial Society of which my host is President.   It
was a weary sitting of five hours with innumerable
speeches.   Of course I had to say "a few words," and
if I can get a copy of the papers I will send them to you.
I flatter myself they were words of wisdom, though hardly
likely to contribute to my popularity among the fishermen.

On the 21st he gave an address on the Herring.
To describe the characteristics of this fish in the
Eastern Counties, he says, might seem like carrying
coals to Newcastle; nevertheless the fisherman's
knowledge is not the same as that of the man of
science, and includes none but the vaguest notions of
the ways of life of the fish and the singularities of its
organisation which perplexed biologists.   His own
study of the problems connected with the herring
had begun nineteen years before, when he served on
the first of his two Fishery Commissions; and one of
his chief objects in this address was to insist upon a
fact, borne out partly by the inquiries of the Commission, partly by later investigations in Europe and
America, which it was difficult to make people

appreciate, namely, the impossibility of man's fisheries affecting the numbers of the herring to any appreciable extent, a year's catch not amounting to the estimated number of a single shoal; while the flatfish and cod fisheries remove many of the most destructive enemies of the herring. Those who had not studied the question in this light would say that "it stands to reason" that vast fisheries must tend to exterminate the fish; apropos of which, he made his well-known remark, that in questions of biology "if any one tells me 'it stands to reason' that such and such things must happen, I generally find reason to doubt the safety of his standing."

This year, also, he began the investigations which completed former inquiries into the subject, and finally elucidated the nature of the salmon disease. The last link in the chain of evidence which proved its identity with a fungoid disease of flies, was not reached until March 1883; and on July 3 following he delivered a full account of the disease, its nature and origin, in an address at the Fisheries Exhibition in London.

In 1881, then, at the end of December, he went to North Wales to study on the fresh fish the nature of the epidemic of salmon disease which had broken out in the Conway, in spite of being in such bad health that he was persuaded to let his younger son come and look after him. But this was only a passing premonition of the breakdown which was to come upon him three years after.

One year's work as Inspector was very like another.
In 1882, for instance, on January 21, he is at Berwick,
"voiceless but jolly"; in the spring he had to attend
a Fisheries Exhibition in Edinburgh, and writes:—

April 12.—We have opened our Exhibition, and I
have been standing about looking at the contents until
my back is broken.

April 13.—The weather here is villainous—a regular
Edinburgh "coorse day." I have seen all I wanted to
see of the Exhibition, eaten two heavy dinners, one with
Primrose and one with Young, and want to get home.
Walpole and I are dining domestically at home this
evening, having virtuously refused all invitations.

In June he was in Hampshire; on July 25 he
writes from Tynemouth :—

I reached here about 5 o'clock, and found the bailiff
or whatever they call him of the Board of Conservators,
awaiting me with a boat at my disposal. So we went off
to look at what they call "The Playground"—two bays
in which the salmon coming from the sea rest and disport
themselves until a fresh comes down the river and they
find it convenient to ascend. Harbottle bailiff in question
is greatly disturbed at the amount of poaching that goes
on in the playground, and unfolded his griefs to me at
length. It was a lovely evening, very calm, and I
enjoyed my boat expedition. To-morrow there is to be
another to see the operations of a steam trawler, which
in all probability I shall not enjoy so much. I shall
take a light breakfast.

These were the pleasanter parts of the work. The
less pleasant was sitting all day in a crowded court,
hearing a disputed case of fishing rights, or examining

witnesses who stuck firmly to views about fish which had long been exploded by careful observation. But on the whole he enjoyed it, although it took him away from research in other departments. This summer, on the death of Professor Rolleston, he was sounded on the question whether he would consent to accept the Linacre Professorship of Physiology at Oxford. He wrote to the Warden of Merton :—

4 MARLBOROUGH PLACE,
June 22, 1881.

MY DEAR BRODRICK—Many thanks for your letter. I can give you my reply at once, as my attention has already been called to the question you ask ; and it is that I do not see my way to leaving London for Oxford. My reasons for arriving at this conclusion are various. I am getting old, and you should have a man in full vigour.   I doubt whether the psychical atmosphere of Oxford would suit me, and still more, whether I. should suit it after a life spent in the absolute freedom of London. And last, but by no means least, for a man with five children to launch into the world, the change would involve a most serious loss of income.   No doubt there are great attractions on the other side ; and, if I had been ten years younger, I should have been sorely tempted to go to Oxford, if the University would have had me.   But things being as they are, I do not see my way to any other conclusion than that which I have reached.

The same feeling finds expression in a letter to Professor (afterwards Sir William) Flower, who was also approached on the same subject, and similarly determined to remain in London.

July 21, 1881.

MY DEAR FLOWER—I am by no means surprised, and except for the sake of the University, not sorry that you have renounced the Linacre.

Life is like walking along a crowded street—there always seem to be fewer obstacles to getting along on the opposite pavement—and yet, if one crosses over, matters are rarely mended.

I assure you it is a great comfort to me to think that you will stay in London and help in keeping things straight in this world of crookedness.

I have thought a good deal about ——, but it would never do. No one could value his excellent qualities of all kinds, and real genius in some directions, more than I do; but, in my judgment, nobody could be less fitted to do the work which ought to be done in Oxford—I mean to give biological science a status in the eyes of the Dons, and to force them to acknowledge it as a part of general education. Moreover, his knowledge, vast and minute as it is in some directions, is very imperfect in others, and the attempt to qualify himself for the post would take him away from the investigations, which are his delight and for which he is specially fitted. . . .

I was very much interested in your account of the poor dear Dean's illness. I called on Thursday morning, meeting Jowett and Grove at the door, and we went in and heard such an account of his state that I had hopes he might pull through. We shall not see his like again.

The last time I had a long talk with him was about the proposal to bury George Eliot in the Abbey, and a curious revelation of the extraordinary catholicity and undaunted courage of the man it was. He would have done it had it been pressed upon him by a strong representation.

I see he is to be buried on Monday, and I suppose and hope I shall have the opportunity of attending.— Ever yours very faithfully,   T. H. HUXLEY.

This letter refers to the death of his old friend
Dean Stanley.   The Dean had long kept in touch
with the leaders of scientific thought, and it is deeply
interesting to know that on her death-bed, five years
before, his wife said to him as one of her parting
counsels, "Do not lose sight of the men of science,
and do not let them lose sight of you."   "And then,"
writes Stanley to Tyndall, "she named yourself and
Huxley."

Strangely enough, the death of the Dean involved
another invitation to Huxley to quit London for
Oxford.   By the appointment of Dean Bradley to
Westminster, the Mastership of University College
was left vacant.   Huxley, who was so far connected
with the college that he had examined there for a
science Fellowship, was asked if he would accept it,
but after careful consideration declined.   He writes
to his son, who had heard rumours of the affair in
Oxford :—

4 MARLBOROUGH PLACE, *Nov.* 4, 1881.

MY DEAR LENS—There is truth in the rumour; in so
far as this that I was asked if I would allow myself to
be nominated for the Mastership of University, that I
took the question into serious consideration and finally
declined.

But I was asked to consider the communication made
to me confidential, and I observed the condition strictly.
The leakage must have taken place among my Oxford
friends, and is their responsibility, but at the same time
I would rather you did not contribute to rumour on the
subject.   Of course I should have told you if I had not
been bound to reticence.

I was greatly tempted for a short time by the prospect of rest, but when I came to look into the matter closely there were many disadvantages. I do not think I am cut out for a Don nor your mother for a Donness—we have had thirty years' freedom in London, and are too old to put in harness.

Moreover, in a monetary sense I should have lost rather than gained.

My astonishment at the proposal was unfeigned, and I begin to think I may yet be a Bishop.—Ever your loving father,   T. H. HUXLEY.

His other occupations this year were the Medical Acts Commission, which sat until the following year, and the International Medical Congress.

The Congress detained him in London this summer later than usual. It lasted from the 3rd to the 9th of August, on which day he delivered a concluding address on "The Connection of the Biological Sciences with Medicine" (*Coll. Ess.* iii. p. 347). He showed how medicine was gradually raised from mere empiricism and based upon true pathological principles, through the independent growth of physiological knowledge, and its correlation to chemistry and physics. "It is a peculiarity," he remarks, "of the physical sciences that they are independent in proportion as they are imperfect." Yet "there could be no real science of pathology until the science of physiology had reached a degree of perfection unattained, and indeed unattainable, until quite recent times." Historically speaking, modern physiology, he pointed out, began with

Descartes' attempt to explain bodily phenomena on purely physical principles; but the Cartesian notion of one controlling central mechanism had to give way before the proof of varied activities residing in various tissues, until the cell-theory united something of either view. "The body is a machine of the nature of an army, not that of a watch or of a hydraulic apparatus." On this analogy, diseases are derangements either of the physiological units of the body, or of their co-ordinating machinery: and the future of medicine depends on exact knowledge of these derangements and of the precise alteration of the conditions by the administration of drugs or other treatment, which will redress those derangements without disturbing the rest of the body.

A few extracts from letters to his wife describe his occupation at the Congress, which involved too much "society" for his liking.

August 4.—The Congress began with great *éclat* yesterday, and the latter part of Paget's address was particularly fine. After, there was the lunch at the Pagets' with the two Royalties. After that, an address by Virchow. After that, dinner at Sanderson's, with a confused splutter of German to the neighbours on my right. After that a tremendous soirée at South Kensington, from which I escaped as soon as I could, and got home at midnight. There is a confounded Lord Mayor's dinner this evening ("the usual turtle and speeches to the infinite bewilderment and delight of the foreigners," August 6), and to-morrow a dinner at the Physiological Society. But I have got off the Kew party, and mean to go quietly down to the Spottiswoodes [*i.e.*

at Sevenoaks] on Saturday afternoon, and get out of the
way of everything except the College of Surgeons' Soirée,
till Tuesday.   Commend me for my prudence.

On the 5th he was busy all day with Government
Committees, only returning to correct proofs of his
address before the social functions of the evening.
Next morning he writes :—

> I have been toiling at my address this morning.   It
> is all printed, but I must turn it inside out, and make a
> speech of it if I am to make any impression on the
> audience in St. James' Hall.   Confound all such
> bobberies.
>
> *August 9.*—I got through my address to-day as well
> as I ever did anything.   There was a large audience, as
> it was the final meeting of the Congress, and to my
> surprise I found myself in excellent voice and vigour.   So
> there is life in the old dog yet.   But I am greatly relieved
> it is over, as I have been getting rather shaky.

When the Medical Congress was over, he joined
his family at Grasmere for the rest of August.   In
September he attended the British Association at
York, where he read a paper on the "Rise and
Progress of Paleontology," and ended the month with
fishery business at Aberystwith and Carmarthen.

The above paper is to be found in *Collected Essays*,
iv. p. 24.   In it he concludes an historical survey of
the views held about fossils by a comparison of the
opposite hypothesis upon which the vast store of
recently accumulated facts may be interpreted ; and
declaring for the hypothesis of evolution, repeats
the remarkable words of the "Coming of Age of the

Origin of Species," that "the paleontological discoveries of the last decade are so completely in
accordance with the requirements of this hypothesis
that, if it had not existed, the paleontologist would
have had to invent it."

In February died Thomas Carlyle. Mention has
already been made of the influence of his writings
upon Huxley in strengthening and fixing once for all,
at the very outset of his career, that hatred of shams
and love of veracity, which were to be the chief
principle of his whole life. It was an obligation he
never forgot, and for this, if for nothing else, he was
ready to join in a memorial to the man. In reply to
a request for his support in so doing, he wrote to
Lord Stanley of Alderley on March 9 :—

> Anything I can do to help in raising a memorial to
> Carlyle shall be most willingly done. Few men can have
> dissented more strongly from his way of looking at things
> than I ; but I should not yield to the most devoted of
> his followers in gratitude for the bracing wholesome
> influence of his writings when, as a very young man, I
> was essaying without rudder or compass to strike out a
> course for myself.

Mention has already been made (p. 302) of his ill-
health at the end of the year, which was perhaps a
premonition of the breakdown of 1883. An indication of the same kind may be found in the following
letter to Mrs. Tyndall, who had forwarded a
document which Dr. Tyndall had meant to send
himself with an explanatory note.

4 Marlborough Place, *March* 25, 1881.

My dear Mrs. Tyndall—But where is his last note to me? That is the question on which I have been anxiously hoping for light since I received yours and the enclosure, which contains such a very sensible proposition that I should like to know how it came into existence, abiogenetically or otherwise.

As I am by way of forgetting everything myself just now, it is a comfort to me to believe that Tyndall has forgotten he forgot to send the letter of which he forgot the inclosure. The force of disremembering could no further go.—In affectionate bewilderment, ever yours,

T. H. Huxley.

His general view of his health, however, was much more optimistic, as appears from a letter to Mrs. May (wife of the friend of his boyhood) about her son, whose strength had been sapped by typhoid fever, and who had gone out to the Cape to recruit.

4 Marlborough Place, *June* 10, 1881.

My dear Mrs. May—I promised your daughter the other day that I would send you the Bishop of Natal's letter to me. Unfortunately I had mislaid it, and it only turned up just now when I was making one of my periodical clearances in the chaos of papers that accumulates on my table.

You will be pleased to see how fully the good Bishop appreciates Stuart's excellent qualities, and as to the physical part of the business, though it is sad enough that a young man should be impeded in this way, I think you should be hopeful. Delicate young people often turn out strong old people—I was a thread paper of a boy myself, and now I am an extremely tough old personage. . . .

With our united kind regards to Mr. May and yourself
—Ever yours very faithfully,   T. H. HUXLEY.

Perhaps if he had been able each year to carry
out the wish expressed in the following letter, which
covered an introduction to Dr. Tyndall at his house
on the Bel Alp, the breakdown of 1883 might have
been averted.

> 4 MARLBOROUGH PLACE, LONDON, N.W.
> *July* 5 (1881 ?)
>
> MY DEAR SKELTON—It is a great deal more than I would
> say for everybody, but I am sure Tyndall will be very
> much obliged to me for making you known to him; and
> if you, insignificant male creature, how very much more
> for the opportunity of knowing Mrs. Skelton!
>
> For which last pretty speech I hope the lady will
> make a prettier curtsey.   So go boldly across the Aletsch,
> and if they have a knocker (which I doubt), knock and it
> shall be opened unto you.
>
> I wish I were going to be there too; but Royal Com-
> missions are a kind of endemic in my constitution, and I
> have a very bad one just now.[1]
>
> With kind remembrances to Mrs. Skelton—Ever yours
> very faithfully,   T. H. HUXLEY.

The ecclesiastical sound of his new title of Dean
of the College of Science afforded him a good deal of
amusement.   He writes from Grasmere, where he had
joined his family for the summer vacation :—

> *Aug.* 18, 1881.
>
> MY DEAR DONNELLY—I am astonished that you don't
> know that a letter to a Dean ought to be addressed " The

---

[1] The Medical Acts Commission, 1881-2.

Very Revd.'   I don't generally stand much upon etiquette, but when my sacred character is touched I draw the line.

We had athletics here yesterday, and as it was a lovely day, all Cumberland and Westmoreland sent contingents to see the fun. . . .

This would be a grand place if it were drier, but the rain it raineth every day—yesterday being the only really fine day since our arrival.

However, we all thrive, so I suppose we are adapting ourselves to the medium, and shall be scaly and finny before long.

Haven't you done with Babylon yet ?   It is high time you were out of it.—Ever yours very faithfully,

T. H. HUXLEY.

CHAPTER XIII

1882

THE year 1882 was a dark year for English science. It was marked by the death of both Charles Darwin and of Francis Balfour, the young investigator, of whom Huxley once said, "He is the only man who can carry out my work." The one was the inevitable end of a great career, in the fulness of time; the other was one of those losses which are the more deplorable as they seem unnecessary, the result of a chance slip, in all the vigour of youth. I remember his coming to our house just before setting out on his fatal visit to Switzerland, and my mother begging him to be careful about risking so valuable a life as his in dangerous ascents. He laughingly replied that he only wanted to conquer one little peak on Mont Blanc. A few days later came the news of his fatal fall upon the precipices of the Aiguille Blanche. Since the death of Edward Forbes, no loss outside the circle of his family had affected my father so deeply. For three days he was utterly prostrated, and was scarcely able either to eat or sleep.

There was indeed a subtle affinity between the two men. My mother, who was greatly attached to Francis Balfour, said once to Sir M. Foster, "He has not got the dash and verve, but otherwise he reminds me curiously of what my husband was in his 'Rattlesnake' days." "How strange," replied Sir Michael, "when he first came to the front, Lankester wrote asking me, 'Who is this man Balfour you are always talking about?' and I answered, 'Well, I can only describe him by saying he is a younger Huxley.'"

Writing to Dr. Dohrn on September 24, Huxley says :—

Heavy blows have fallen upon me this year in losing Darwin and Balfour, the best of the old and the best of the young. I am beginning to feel older than my age myself, and if Balfour had lived I should have cleared out of the way as soon as possible, feeling that the future of Zoological Science in this country was very safe in his hands. As it is, I am afraid I may still be of use for some years, and shall be unable to sing my "Nunc dimittis" with a good conscience.

Darwin was in correspondence with him till quite near the end; having received the volume *Science and Culture*, he wrote on January 12, 1882 :—

With respect to automatism,[1] I wish that you could review yourself in the old, and, of course, forgotten, trenchant style, and then you would have [to] answer yourself with equal incisiveness; and thus, by Jove, you might go on *ad infinitum* to the joy and instruction of the world.

---

[1] The allusion is to the 1874 address on "Animals as Automata," which was reprinted in *Science and Culture*.

And again on March 27 :—

Your most kind letter has been a real cordial to me.
. . . Once again accept my cordial. thanks, my dear old
friend.   I wish to God there were more automata in the
world like you

Darwin died on April 19, and a brief notice being
required for the forthcoming number of *Nature* on
the 27th, Huxley made shift to write a brief article,
which is printed in the *Collected Essays*, ii. p. 244.
But as neither he nor Sir Joseph Hooker could at
the moment undertake a regular obituary notice, this
was entrusted to Professor Romanes, to whom the
following letters were written.

4 MARLBOROUGH PLACE, *April* 26, 1882.

MY DEAR ROMANES—Thank you for your hearty
letter.   I spent many hours over the few paragraphs I
sent to *Nature*, in trying to express what all who
thoroughly knew and therefore loved Darwin, must feel
in language which should be absolutely free from rhetoric
or exaggeration.

I have done my best, and the sad thing is that I
cannot look for those cheery notes he used to send me
in old times, when I had written anything that pleased
him.

In case we should miss one another to-day, let me
say that it is impossible for me to undertake the obituary
in *Nature*.   I have a conglomeration of business of various
kinds upon my hands just now.   I am sure it will be
very safe in your hands.—Ever yours very faithfully,

T. H. HUXLEY.

Pray do what you will with what I have written in
Nature.

4 MARLBOROUGH PLACE, *May* 9, 1882.

MY DEAR ROMANES—I feel it very difficult to offer any useful criticism on what you have written about Darwin, because, although it does not quite please me, I cannot exactly say how I think it might be improved. My own way is to write and re-write things, until by some sort of instinctive process they acquire the condensation and symmetry which satisfies me. And I really could not say how my original drafts are improved until they somehow improve themselves.

Two things however strike me. I think there is too much of the letter about Henslow. I should be disposed to quote only the most characteristic passages.

The other point is that I think strength would be given to your panegyric by a little pruning here and there.

I am not likely to take a low view of Darwin's position in the history of science, but I am disposed to think that Buffon and Lamarck would run him hard in both genius and fertility. In breadth of view and in extent of knowledge these two men were giants, though we are apt to forget their services. Von Bär was another man of the same stamp; Cuvier, in a somewhat lower rank, another; and J. Müller another.

"Colossal" does not seem to me to be the right epithet for Darwin's intellect. He had a clear rapid intelligence, a great memory, a vivid imagination, and what made his greatness was the strict subordination of all these to his love of truth.

But you will be tired of my carping, and you had much better write what seems right and just to yourself. —Ever yours very faithfully,   T. H. HUXLEY.

Two scientific papers published this year were on subjects connected with his work on the fisheries, one "A Contribution to the Pathology of the Epidemic

known as the 'Salmon Disease'" read before the
Royal Society on the occasion of the Prince of Wales
being admitted a Fellow (February 21; *Proc. Roy.
Soc.* xxxiii. pp. 381-389); the other on "Saprolegnia
in relation to the Salmon Disease" (*Quarterly Journal
of Microscopical Science,* xxii. pp. 311-333). A third,
at the Zoological Society, was on the "Respiratory
Organs of Apteryx" (*Proc. Z. S.* 1882, pp. 560-569).
He delivered an address before the Liverpool Institu-
tion on "Science and Art in Relation to Education"
(*Coll. Ess.* iii. p. 160), and was busy with the Medical
Acts Commission, which reported this year.

The aim of this Commission [1] was to level up the
varying qualifications bestowed by nearly a score of
different licensing bodies in the United Kingdom,
and to establish some central control by the State
over the licensing of medical practitioners.

The report recommended the establishment of
Boards in each division of the United Kingdom
containing representatives of all the medical bodies in
the division. These boards would register students,
and admit to a final examination those who had
passed the preliminary and minor examinations at
the various universities and other bodies already
granting degrees and qualifications. Candidates who
passed this final examination would be licensed by
the General Medical Council, a body to be elected

---

[1] For a fuller account of this Commission and the part played
in it by Huxley, see his "State and Medical Education" (*Coll.
Ess.* iii. 323), published 1884.

no longer by the separate bodies interested in medical education, but by the Divisional Boards.

The report rejected a scheme for joint examination by the existing bodies, assisted by outside examiners appointed by a central authority, on the ground of difficulty and expense, as well as one for a separate State examination. It also provided for compensation from the fees to be paid by the candidates to existing bodies whose revenues might suffer from the new scheme.

To this majority report, six of the eleven Commissioners appended separate reports, suggesting other methods for carrying out the desired end. Among the latter was Huxley, who gave his reasons for dissenting from the principle assumed by his colleagues, though he had signed the main report as embodying the best means of carrying out a reform, that principle being granted.

"The State examination," he thought, "was ideally best, but for many reasons impossible." But the "conjoint scheme" recommended in the report appeared to punish the efficient medical authorities for the abuses of the inefficient. Moreover, if the examiners of the Divisional Board did not affiliate themselves to any medical authority, the compensation to be provided would be very heavy; if they did, "either they will affiliate without further examination, which will give them the pretence of a further qualification, without any corresponding reality, or they will affiliate in examination, in which case the new examination deprecated by the general voice of the profession will be added, and any real difference between the plan proposed and the 'State examination' scheme will vanish."

The compensation proposed, too, would chiefly fall to the discredited bodies, who had neglected their duties.

> The scheme (he writes in his report), which I ventured to suggest is of extreme simplicity; and while I cannot but think that it would prove thoroughly efficient, it interferes with no fair vested interest in such a manner as to give a claim for compensation, and it inflicts no burden either in the way of taxation or extra examination on the medical profession.
>
> This proposal is, that if any examining body satisfies the Medical Council (or other State authority), that it requires full and efficient instruction and examination in the three branches of medicine, surgery, and midwifery; and if it admits a certain number of coadjutor examiners appointed by the State authority, the certificate of that authority shall give admission to the Medical Register.
>
> I submit that while the adopting this proposal would secure a practically uniform minimum standard of examination, it would leave free play to the individuality of the various existing or future universities and medical corporations; that the revenues of such bodies derived from medical examinations would thenceforth increase or diminish in the ratio of their deserts; that a really efficient inspection of the examinations would be secured, and that no one could come upon the register without a complete qualification.

That there was no difficulty in this scheme was shown by the experience of the Scotch Universities;

and the expense would be less than the proposed compensation tax.

The chief part of the summer vacation Huxley spent at Lynton, on the north coast of Devonshire. "The Happy Family," he writes to Dr. Dohrn, "has been spending its vacation in this pretty place, eighteen miles of up hill and down dale from any railway." It was a country made for the long rambles he delighted in after the morning's due allowance of writing. And although he generally preferred complete quiet on his holidays, with perfect freedom from all social exigencies, these weeks of rest were rendered all the pleasanter by the unstudied and unexacting friendliness of the family party which centred around Mr. and Mrs. F. Bailey of Lee Abbey hard by—Lady Tenterden, the Julius and the Henry Pollocks, the latter old friends of ours.

Though his holiday was curtailed at either end, he was greatly set up by it, and writes to chaff his son-in-law for taking too little rest—

I was glad to hear that F. had stood his fortnight's holiday so well ; three weeks might have knocked him up !

On the same day, September 26, he wrote the letter to Dr. Dohrn, mentioned above, answering two inquiries—one as to arrangements for exhibiting at the Fisheries Exhibition to be held in London the following year, the other as to whether England would follow the example of Germany and Italy in sending naval officers to the Zoological Station at

Naples to be instructed in catching and preserving marine animals for the purposes of scientific research.

With respect to question No. 2, I am afraid my answer must be less hopeful. So far as the British Admiralty is represented by the ordinary British admiral, the only reply to such a proposition as you make that I should expect would be that he (the British admiral, to wit) would see you d—d first. However, I will speak of the matter to the Hydrographer, who really is interested in science, at the first opportunity.

For many years before this, and until the end of his life, there was another side to his correspondence which deserves mention.

I wish that more of the queer letters, which arrived in never-failing streams, had been preserved. A favourite type was the anonymous letter. It prayed fervently, over four pages, that the Almighty would send him down quick into the pit, and was usually signed simply "A Lady." Others came from cranks of every species: the man who demonstrated that the world was flat, or that the atmosphere had no weight—an easy proof, for you weigh a bottle full of air; then break it to pieces, so that it holds nothing; weigh the pieces, and they are the same weight as the whole bottle full of air! Or, again, that the optical law of equality between the angle of incidence and the angle of reflection is a delusion, whence it follows that all our established latitudes are incorrect, and the difference of temperature between Labrador and Ireland, nominally on the same parallel, is easily accounted for. Then came the suggestions of little pieces of work that might so easily be undertaken by a man of Huxley's capacity, learning, and energy. Enormous manuscripts were sent him with a request that he would write a careful criticism of them, and arrange for their publication in the proceedings of some learned society or first-rate magazine. One of the most delightful came this year. A doctor in India, having just read *John Inglesant*, begged Professor Huxley to do for Science what Mr. Shorthouse had done for the Church of England. As for the material difficulties in the way of getting such a book written in the midst of other work, the ingenious doctor suggested the use of a phonograph driven by a gas-engine. The great thoughts dictated into it from the comfort of an armchair, could easily be worked up into novel shape by a collaborator.

India, again, provided the following application of 1885, made in all seriousness by a youthful Punjaubee with scientific aspirations, who feared to be forced into the law. After an intimate account of his life, he modestly appeals for a post in some scientific institution, where he may get his food, do experiments three or four hours a day, and learn English. Latterly his mental activity had been very great:—"I have been contemplating," he says, "to give a new system of Political Economy to the world. I have questioned, perhaps with success, the

validity of some of the fundamental doctrines of
H. Spencer's synthetic philosophy," and so on.

Another remarkable communication is a reply-
paid telegram from the States, in 1892, which ran as
follows :—

> Unless all reason and all nature have deceived me, I
> have found the truth. It is my intention to cross the
> ocean to consult with those who have helped me most to
> find it. Shall I be welcome? Please answer at my
> expense, and God grant we all meet in life on earth.

Another, of British origin this time, was from a
man who had to read a paper before a local Literary
Society on the momentous question, "Where are
we?" so he sent round a circular to various
authorities to reinforce his own opinions on the six
heads into which he proposed to divide his discourse, viz. :

Where are we in Space?
„   „   Science?
„   „   Politics?
„   „   Commerce?
„   „   Sociology?
„   „   Theology?

The writer received an answer, and a mild one:

> Any adequate reply to your inquiry would be of the
> nature of a treatise, and that, I regret, I cannot under-
> take to write.

Two letters of this year touch on Irish affairs, in
which he was always interested, having withal a
certain first-hand knowledge of the people and the

country they lived in, from his visits there, both as a
Fishery Commissioner and on other occasions. He
writes warmly to the historian who treated of Ireland
without prejudice or rancour.

4 MARLBOROUGH PLACE,
April 16, 1882.

MY DEAR LECKY—Accept my best thanks for your
two volumes, which I found on my return from Scotland
yesterday.

I can give no better evidence of my appreciation of
their contents than by the confession that they have
caused me to neglect my proper business all yesterday
evening and all to-day.

The section devoted to Irish affairs is a model of
lucidity, and bears on its face the stamp of justice and
fair dealing. It is a most worthy continuation of the
chapter on the same subject in the first volume, and that
is giving high praise.

You see I write as if I knew something about the
subject, but you are responsible for creating the
delusion.

With kindest remembrances to Mrs. Lecky—Ever
yours very faithfully,

T. H. HUXLEY.

A few weeks later, the murder of Lord Frederick
Cavendish sent a thrill of horror throughout England.
Huxley was as deeply moved as any, but wrote calmly
of the situation.

To his Eldest Son

4 MARLBOROUGH PLACE,
May 9, 1882.

MY DEAR LEONARD—Best thanks for your good wishes.[1] Notwithstanding the disease of A.D., which always proves mortal sooner or later, I am in excellent case. . . .

I knew both Lord F. Cavendish and his wife and Mr. Burke. I have never been able to get poor Lady Frederick out of my head since the news arrived.

The public mind has been more stirred than by anything since the Indian Mutiny. But if the Government keep their heads cool, great good may come out of the evil, horrible as it is. The Fenians have reckoned on creating an irreparable breach between England and Ireland. It should be our business to disappoint them first and extirpate them afterwards. But the newspaper writers make me sick, especially the *Times.*—Ever your affectionate father,   T. H. HUXLEY.

It is interesting, also, to see how he appeared about this time to one of a younger generation, acute, indeed, and discriminating, but predisposed by circumstances and upbringing to regard him at first with curiosity rather than sympathy. For this account I am indebted to one who has the habit, so laudable in good hands, of keeping a journal of events and conversations. I have every confidence in the substantial accuracy of so well trained a reporter.

---

[1] For his birthday, May 4.

EXTRACT FROM JOURNAL

Nov. 25, 1882.

In the evening we dined at the——'s, chiefly a family party with the addition of Professor Huxley and his wife and ourselves. Much lively conversation, after dinner, begun among the ladies, but continued after the gentlemen appeared, on the subjects of Truth, Education, and Women's Rights, or, more strictly speaking, women's capabilities. Our hostess (Lady——) was, if possible, more vehement and paradoxical than her wont, and vigorously maintained that *truth* was no virtue in itself, but must be inculcated for expediency's sake. The opposite view found a champion in Professor Huxley, who described himself as "almost a fanatic for the sanctity of truth." Lady——urged that truth was often a very selfish virtue, and that a man of noble and unselfish character might lie for the sake of a friend, to which some one replied that after a course of this unselfish lying the noble character was pretty sure to deteriorate, while the Professor laughingly suggested that the owner had a good chance of finding himself landed ultimately in Botany Bay.

The celebrated instance of John Inglesant's perjury for the sake of Charles I. was then brought forward, and it was this which led Professor Huxley to say that in his judgment no one had the right passively to submit to a false accusation, and that "moral suicide" was as blameworthy as physical suicide. "He may refuse to commit another, but he ought not to allow himself to be believed worse than he actually is. It is a loss to the world of *moral force*, which cannot be afforded."

. . . Then as regards women's powers. The Professor said he did not believe in their ever succeeding in a competition with men. Then he went on:—"I can't help looking at women with something of the eye of a physiologist. Twenty years ago I thought the womanhood of England was going to the dogs," but now, he said, he observed a wonderful change for the better.   We asked to what he attributed it.   Was it to lawn tennis and the greater variety of bodily exercises ?   " Partly," he answered, " but much more to their having more *pursuits*—more to interest them and to occupy their thoughts and time."

The following letter bears upon the question of employing retired engineer officers in administrative posts in the Science and Art Department :—

THE ROOKERY, LYNTON,
Sept. 19, 1882.

MY DEAR DONNELLY—Your letter seems to have arrived here the very day I left for Whitby, whither I had to betake myself to inspect a weir, so I did not get it until my return last night.

I am extremely sorry to hear of the possibility of Martin's giving up his post.   He took so much interest in the work and was so very pleasant to deal with, that I do not think we shall easily find any one to replace him.

If you will find another R.E. at all like him, in Heaven's name catch him and put him in, job or no job.

The objection to a small clerk is that we want somebody who knows how to deal with men, and especially young men on the one hand, and especially cantankerous (more or less) old scientific buffers on the other.

The objection to a man of science is that (1) we want a man of business and not a m.s., and (2) that no man scientifically worth having that I know of is likely to take such an office.

"As at present advised" I am all for an R.E., so I cannot have the pleasure even of trying to convert you.

With our united kindest regards—Ever yours very faithfully,   T. H. HUXLEY.

I return next Monday.

Two letters of thanks follow, one at the beginning of the year to Mr. Herbert Spencer for the gift of a very fine photograph of himself; the other, at the end of the year, to Mr. (afterwards Sir John) Skelton, for his book on Mary Queen of Scots and the Casket Letters.

As to the former, it must be premised that Mr. Spencer abhorred exaggeration and inexact talk, and would ruthlessly prick the airy bubbles which endued the conversation of the daughters of the house with more buoyancy than strict logic, a gift which, he averred, was denied to woman.

4 MARLBOROUGH PLACE,
Jan. 25, 1882.

MY DEAR SPENCER—Best thanks for the photograph. It is very good, though there is just a touch of severity in the eye. We shall hang it up in the dining-room, and if anybody is guilty of exaggerated expressions or bad logic (five womenkind habitually sit round that table), I trust they will feel that that eye is upon them. —Ever yours very faithfully,

T. H. HUXLEY

4 MARLBOROUGH PLACE, N.W
Dec. 31, 1882.

MY DEAR SKELTON—If I may not thank you for the book you have been kind enough to send me, I may at anyrate wish you and Mrs. Skelton a happy New Year and many on 'em.

I am going to read your vindication of Mary Stuart as soon as I can. Hitherto I am sorry to say I have classed her with Eve, Helen, Cleopatra, Delilah, and sundry other glorious ——s who have lured men to their destruction.

But I am open to conviction, and ready to believe that she blew up her husband only a little more thoroughly than other women do, by reason of her keener perception of logic.—Ever yours very faithfully,

T. H. HUXLEY.

CHAPTER XIV

1883

THE pressure of official work, which had been constantly growing since 1880, reached its highest point in 1883. Only one scientific memoir[1] was published by him this year, and then no more for the next four years. The intervals of lecturing and examining were chiefly filled by fishery business, from which, according to his usual custom when immersed in any investigation, he chose the subject, "Oysters and the Oyster Question," both for his Friday evening discourse at the Royal Institution on May 11, and for his course to Working Men between Jan. 8 and Feb. 12.

There are the usual notes of all seasons at all parts of England. A deserted hotel at Cromer in January was uninviting.

My windows look out on a wintry sea, and it is bitter cold. Notwithstanding, a large number of the aquatic

---

[1] Contributions to Morphology, Ichthyopsida, No. 2. On the Oviducts of Osmerus ; with remarks on the relations of the Teleostean with the Ganoid Fishes (*Proc. Zool. Soc.* 1883, pp. 132-139).

gentlemen to whom I shall have the pleasure of listening, by and by, are loafing against the railings opposite, as only fishermen can loaf.

In April he had been ill, and his wife begged him to put off some business which had to be done at York. But unless absolutely ordered to bed by his doctor, nothing would induce him to put personal convenience before public duty. However, he took his son to look after him.

> I am none the worse for my journey (he writes from York), rather the better; so Clark is justified, and I should have failed in my duty if I had not come. H. looks after me almost as well as you could do.

To make amends, fishery business in the west country during a fine summer had "a good deal of holiday in it," through a cross journey at the beginning of August from Abergavenny to Totness made him write :—

> If ever (except to-morrow, by the way) I travel within measurable distance of a Bank Holiday by the Great Western, may jackasses sit on my grandmother's grave.

As the business connected with the Inspectorship had been enlarged in the preceding years by exhibitions at Norwich and Edinburgh, so it was enlarged this year, and to a still greater extent, by the Fisheries Exhibition in London. This involved upon him as Commissioner, not only the organisation of the Conference on Fish Diseases and the paper on the Diseases of Fish already mentioned, but administration, committee meetings, and more—a speech on behalf of the Commissioners in reply to the welcome given them by the Prince of Wales at the opening of the Exhibition. On the following day he expressed his feelings at this mode of spending his time in a letter to Sir M. Foster.

I am dog-tired with yesterday's function. Had to be at the Exhibition in full fig at 10 A.M., and did not get home from the Fishmongers' dinner till 1.20 this morning.

Will you tell me what all this has to do with my business in life, and why the last fragments of a misspent life that are left to me are to be frittered away in all this drivel ?—Yours savagely,

T. H. H.

Later in the year, also, he had to serve on another Fishery Commission much against his will, though on the understanding that, in view of his other engagements, he need not attend all the sittings.

A more satisfactory result of the Exhibition was that he found himself brought into close contact with several of the great city companies, whose enormous resources he had long been trying, not without some success, to enlist on behalf of technical and scientific education.

Among these may be noted the Fishmongers, the Mercers, who had already interested themselves in technical education, and gave their hall for the meetings of the City and Guilds Council, of which

Huxley was an active member; the Clothworkers, in whose schools he distributed the prizes this year; and, not least, the Salters, who presented him with their freedom on November 13. Their master, Mr. J. W. Clark, writing in August, after Huxley had accepted their proposal, says: "I think you must admit that the City Companies have yielded liberally to the gentle compression you have exercised on them. So far from helping you to act the traitor, we propose to legitimise your claim for education, which several of us shall be willing to unite with you in promoting" (see pp. 219, 220).

The crowning addition, however, to Huxley's official work was the Presidency of the Royal Society. He had resigned the Secretaryship in 1880, after holding office for nine years under three Presidents —Airy, Hooker, and Spottiswoode. Spottiswoode, like Hooker, was a member of the x Club, and was regarded with great affection and respect by Huxley, who in 1887 wrote of him to Mr. John Morley:—

It is quite absurd you don't know Spottiswoode, and I shall do both him and you a good turn by bringing you together. He is one of my best friends, and comes under the A1 class of "people with whom you may go tiger-hunting."

On June 7, writing to Professor (afterwards Sir E.) Frankland, he says:—

You will have heard that Spottiswoode is seriously ill. The physicians suspect typhoid, but are not quite certain. I called this morning, and hear that he remains much as

he has been for the last two or three days. So many of
our friends have dropped away in the course of the last
two years that I am perhaps morbidly anxious about
Spottiswoode, but there is no question that his condition
is such as to cause grave anxiety.

But by the end of the month his fears were
realised. Consequently it devolved upon the Council
of the Royal Society to elect one of their own body
to hold office until the St. Andrew's Day following,
when a regular President would be elected at a
general meeting of the Society.

Huxley himself had no wish to stand. He writes to
Sir M. Foster on June 27, announcing Spottiswoode's
death, which had taken place that morning :—

It is very grievous in all ways. Only the other day
he and I were talking of the almost miraculous way in
which the *x* Club had held together without a break for
some 18 years, and little did either of us suspect that he
would be the first to go.

A heavy responsibility falls on you in the Royal
Society. It strikes me you will have to call another
meeting of the Council before the recess for the consideration of the question of the Presidency. It is hateful to
talk of these things, but I want you to form some notion
of what had best be done as you come up to-morrow.

——— is a possibility, but none of the other officers, I
think.

Indeed, he wished to diminish his official distractions rather than to increase them. His health was
unlikely to stand any additional strain, and he longed
to devote the remainder of his working years to his

unfinished scientific researches.   But he felt very
strongly that the President of the Royal Society
ought to be chosen for his eminence in science, not
on account of social position, or of wealth, even
though the wealth might have been acquired through
the applications of science.   The acknowledgment
of this principle had led some years back to the
great revolution from within, which succeeded in
making the Society the living centre and representative of science for the whole country, and he was
above all things anxious that the principle should be
maintained.   He was assured, however, from several
quarters that unless he allowed himself to be put
forward, there was danger lest the principle should
be disregarded.

Moved by these considerations of public necessity,
he unwillingly consented to be nominated, but only
to fill the vacancy till the general meeting, when the
whole Society could make a new choice.   Yet even
this limitation seemed difficult to maintain in the
face of the widely expressed desire that he would
then stand for the usual period of five years.   "The
worst of it is," he wrote to Sir M. Foster on July 2,
"that I see myself gravitating towards the Presidency
en permanence, that is to say, for the ordinary period.
And that is what I by no means desired.   —— has
been at me (as a sort of deputation, he told me, from
a lot of the younger men) to stand.   However, I
suppose there is no need to come to any decision
yet."

The following letters, in reply to congratulations on his election, illustrate his attitude of mind in the affair :—

To the Warden of Merton

HINDHEAD, *July* 8, 1883.

MY DEAR BRODRICK—I do not get so many pleasant letters that I can afford to leave the senders of such things unthanked.

I am very much obliged for your congratulations, and I may say that I accepted the office *inter alia* for the purpose of getting people to believe that such places may be properly held by people who have neither riches nor station—who want nothing that statesmen can give—and who care for nothing except upholding the dignity and the freedom of science.—Ever yours very faithfully,

T. H. HUXLEY.

To Sir W. H. Flower, F.R.S.

4 MARLBOROUGH PLACE, *July* 7, 1883.

MY DEAR FLOWER—I am overwhelmed by the kind letters I get from all sides, and I need hardly say that I particularly value yours.

A month ago I said that I ought not, could not, and would not take the Presidency under any circumstances whatever. My wife was dead against it, and you know how hen-pecked I am.

Even when I was asked to take the Presidency to the end of the year and agreed, I stipulated for my freedom next St. Andrew's Day.

But such strong representations were made to me by some of the younger men about the dangers of the situation, that at the last moment almost I changed my mind.

However, I wanted it to be clearly understood that

the Council and the Society are, so far as I am concerned,
perfectly free to put somebody else in my place next
November. All I stipulate for is that my successor shall
be a man of science.

I will not, if I can help it, allow the chair of the
Royal Society to become the appanage of rich men, or have
the noble old Society exploited by enterprising commercial
gents who make their profit out of the application of
science.

Mrs. President was *not* pleased—quite the contrary—
but she is mollified by the kindly expressions, public and
private, which have received the election.

And there are none which we both value more than
yours. (I see I said that before, but I can't say it too
often.)—Ever yours very faithfully,

T. H. HUXLEY.

HINDHEAD, July 8, 1883.

MY DEAR FLOWER—Many thanks for your comforting
letter. When I am fairly committed to anything I
generally have a cold fit—and your judgment that I
have done right is "grateful and comforting" like Epps'
Cocoa. It is not so much work as distraction that is
involved; and though it may put a stop to my purely
scientific work for a while, I don't know that I could be
better employed in the interests of science than in trying
to keep the Royal Society straight.

My wife was very much against it at first—and indeed
when I was first spoken to I declared that I would not
go on after next St. Andrew's Day. But a good deal of
pressure was brought to bear by some of my friends, and
if the Fellows don't turn me out I shall say with
MacMahon, "J'y suis et j'y reste."—Ever yours very
faithfully   T. H. HUXLEY.

We have run down here for a day, but are back
to-morrow.

4 MARLBOROUGH PLACE, *July* 10, 1883.

MY DEAR SPENCER—What an agreeable surprise your letter has been. I have been expecting the most awful scolding for taking more work, and behold as sweetly congratulatory an epistle as a man could wish.

Three weeks ago I swore by all my gods that I would not take the offer at any price, but I suppose the infusion of Theism was too homœopathic for the oath to bind.

Go on sleeping, my dear friend. If you are so amiable with three nights, what will you be with three weeks?

What a shame no rain is sent you. You will be speaking about Providence as I heard of a Yankee doing the other day—" Wal, sir, I guess he's good ; but he's careless."

I think there is a good deal in that view of the government of the world.—Ever yours very faithfully,

T. H. HUXLEY.

TO HIS ELDEST DAUGHTER

4 MARLBOROUGH PLACE, *July* 14, 1883.

DEAREST JESS — I am not sure either whether my accession to the Presidency is a matter for congratulation. Honour and glory are all very fine, but on the whole I prefer peace and quietness, and three weeks ago I declared I would have nothing to do with it.

But there are a good many circumstances in the present state of affairs which weighed heavily in the scale, and so I made up my mind to try the experiment.

If I don't suit the office or the office don't suit me, there is a way out every 30th of November.

There was more work connected with the Secretary-ship—but there is more trouble and responsibility and distraction in the Presidency.

I am amused with your account of your way of

governing your headstrong boy. I find the way of governing headstrong men to be very similar, and I believe it is by practising the method that I get the measure of success with which people credit me.

But they are often very fractious, and it is a bother for a man who was meant for a student.

Poor Spottiswoode's death was a great blow to me. Never was a better man, and I hoped he would stop where he was for the next ten years. . . .—Ever your loving father,   T. H. HUXLEY

He finally decided that the question of standing again in November must depend on whether this course was likely to cause division in the ranks of the Society. He earnestly desired to avoid anything like a contest for scientific honours; [1] he was almost morbidly anxious that the temporary choice of himself should not be interpreted as binding the electors in any way.

I give the following letters to show his sensitiveness on every question of honour and of public advantage :—

BRECHIN CASTLE, BRECHIN, N.B.,
Sept. 19, 1883.

MY DEAR FOSTER—We got here yesterday. The Commission does not meet till next week, so like the historical donkey of Jeshurun I have nothing to do but to wax fat and kick in this excellent pasture.

---

[1] As he wrote a little later :—"I have never competed in the way of honour in my life, and I cannot allow myself to be even thought of as in such a position now, where, with all respect to the honour and glory, they do not appear to me to be in any way equivalent to the burden. And I am not at all sure that I may not be able to serve the right cause outside the Chair rather than in it."

At odd times lately my mind has been a good deal exercised about the Royal Society. I am quite willing to go on in the Chair if the Council and the Society wish it. But it is quite possible that the Council who chose me when the choice was limited to their own body, might be disposed to select some one else when the range of choice is extended to the whole body of the Society. And I am very anxious that the Council should be made to understand, when the question comes forward for discussion after the recess, that the fact of present tenancy constitutes no claim in my eyes.

The difficulty is, how is this to be done? I cannot ask the Council to do as they please, without reference to me, because I am bound to assume that that is what they will do, and it would be an impertinence to assume the contrary.

On the other hand, I should at once decline to be put in nomination again, if it could be said that by doing so I had practically forced myself either upon the Council or upon the Society.

Heaven be praised I have not many enemies, but the two or three with whom I have to reckon don't stick at trifles, and I should not like by any inadvertence to give them a handle.

I have had some thought of writing a letter to Evans,[1] such as he could read to the Council at the first meeting in October, at which I need not be present.

The subject could then be freely discussed, without any voting or resolution on the minutes, and the officers could let me know whether in their judgment it is expedient I should be nominated or not.

In the last case I should withdraw on the ground of my other occupations—which, in fact, is a very real obstacle, and one which looms large in my fits of blue-

---

[1] Sir John Evans, K.C.B., then Treasurer of the Royal Society.

devils, which have been more frequent of late than they should be in holiday time.

Now, will you turn all this over in your mind? Perhaps you might talk it over with Stokes.

Of course I am very sensible of the honour of being P.R.S., but I should be much more sensible of the dishonour of being in that place by a fluke, or in any other way, than by the free choice of the Council and Society.

In fact I am inclined to think that I am morbidly sensitive on the last point; and so, instead of acting on my own impulse, as I have been tempted to do, I submit myself to your worship's wisdom.

I am not sure that I should not have been wiser if I had stuck to my original intention of holding office only till St. Andrew's Day.—Ever yours very faithfully,

T. H. HUXLEY.

SECRETARY OF STATE, HOME DEPARTMENT,
Oct. 3, 1883.

MY DEAR FOSTER—There was an Irish bricklayer who once bet a hodman he would not carry him up to the top of an exceeding high ladder in his hod.   The hodman did it, but Paddy said, " I had great hopes, now, ye'd let me fall just about six rounds from the top."

I told the story before when I was up for the School Board, but it is so applicable to the present case that I can't help coming out with it again.

If you, dear good hodmen, would have but let me fall !

However, as the thing is to be, it is very pleasant to find Evans and Williamson and you so hearty in the process of elevation, and in spite of blue-devils I will do my best to " do my duty in the state of life I'm called to."

But I believe you never had the advantage of learning the Church Catechism.

If there is any good in what is done you certainly deserve the credit of it, for nothing but your letter stopped me from kicking over the traces at once. Do you see how Evolution is getting made into a bolus and oiled outside for the ecclesiastical swallow ? [1]—Ever thine,

THOMAS, P.R.S.

The same feeling appears in his anxiety as President to avoid the slightest appearance of committing the Society to debatable opinions which he supported as a private individual. Thus, although he had "personally, politically, and philosophically" no liking for Charles Bradlaugh, he objected on general grounds to the exclusion of Mrs. Besant and Miss Bradlaugh from the classes at University College, and had signed a memorial in their favour. On the other hand, he did not wish it to be asserted that the Royal Society, through its President, had thrown its influence into what was really a social and political, not a scientific question. He writes to Sir M. Foster on July 18 :—

It is very unlucky for me that I signed the memorial requesting the Council of University College to reconsider their decision about Mrs. Besant and Miss Bradlaugh when I was quite innocent of any possibility of holding the P.R.S.

I must go to the meeting of members to-day and define my position in the matter with more care, under the circumstances.

Mrs. Besant was a student in my teacher's class here last year, and a very well-conducted lady-like person ; but I have never been able to get hold of the "Fruits of

---

[1] This refers to papers read before the Church Congress that year by Messrs. W. H. Flower and F. Le Gros Clarke.

Philosophy," and do not know to what doctrine she has committed herself.

They seem to have excluded Miss Bradlaugh simply on the *noscitur a sociis* principle.

It will need all the dexterity I possess to stand up for the principle of religious and philosophical freedom, without giving other people a hold for saying that I have identified myself with Bradlaugh.

It was the same a little later with the Sunday Society, which had offered him its presidency. He writes to the Hon. Sec. on Feb. 11, 1884 :—

I regret that it is impossible for me to accept the office which the Sunday Society honours me by offering.

It is not merely a disinclination to add to the work which already falls to my share which leads me to say this. So long as I am President of the Royal Society, I shall feel bound to abstain from taking any prominent part in public movements as to the propriety of which the opinions of the Fellows of the Society differ widely.

My own opinions on the Sunday question are exactly what they were five-and-twenty years ago. They have not been hid under a bushel, and I should not have accepted my present office if I had felt that so doing debarred me from reiterating them whenever it may be necessary to do so.

But that is a different matter from taking a step which would, in the eyes of the public, commit the Royal Society, through its President, to one side of the controversy in which you are engaged, and in which I, personally, hope you may succeed as warmly as ever I did.

One other piece of work during the first half of the year remains to be mentioned, namely, the Rede Lecture, delivered at Cambridge on June 12. This was a discourse on Evolution, based upon the consideration of the Pearly Nautilus.

He first traced the evolution of the individual from the ovum, and replied to the three usual objections raised to evolution, that it is impossible, immoral, and contrary to the argument of design, by replying to the first, that it does occur in every individual; to the second, that the morality which opposes itself to truth commits suicide; and to the third that Paley—the most interesting Sunday reading allowed him when a boy—had long since answered this objection.

Then he proceeded to discuss the evolution of the 100 species, all extinct but two, of Nautilus. The alternative theory of new construction, a hundred times over, is opposed alike to tradition and to sane science. On the other hand, evolution, tested by paleontology, proves a sound hypothesis. The great difficulty of science is in tracing every event to those causes which are in present operation; the hypothesis of evolution is analogous to what is going on now.

The summer was passed at Milford, near Godalming, in a house at the very edge of the heather country which from there stretches unbroken past Hindhead and into Wolmer Forest. So well did he like the place that he took it again the following year. But his holiday was like to have been spoilt at the beginning by the strain of an absurd misadventure which involved much fatigue and more anxiety.

I came back only last night (he writes to Sir M. Foster on August 1) from Paris, where I sped on Sunday night, in a horrid state of alarm from a cursed blundering telegram which led me to believe that Leonard (you know he got his first class to our great joy) who had left for the continent on Saturday, was ill or had had an accident.

It was indeed a hurried journey. On receipt of the telegram, he rushed to Victoria only to miss the night mail. The booking-clerk suggested that he should drive to London Bridge, take train to Lewes, and thence take a fly to Newhaven, where he ought to catch a later boat. The problem was to catch the London Bridge train. There was barely a quarter of an hour, but thanks to a good horse and the Sunday absence of traffic, the thing was done, establishing, I believe, what the modern mind delights in, a record in cab-driving. Happily the anxiety at not finding his son in Paris was soon allayed by another telegram from home, where his son-in-law, the innocent sender of the original message, had meanwhile arrived. He writes to Sir M. Foster :—

Judging by my scrawl, which is worse than usual, I should say the anxiety had left its mark, but I am none the worse otherwise.

This was indeed the case. Other letters to Sir M. Foster show that he was unusually well, perhaps because he was really making holiday to some extent. Thus on August 16, he writes :—

This is a lovely country, and I have been reading novels and walking about for the last four days. I must be all right, wind and limb, for I walked over twenty miles the day before yesterday, and except a blister on one heel, was none the worse

And again on September 12 :—

Have been very lazy lately, which means that I have done a great many things that I need not have done, and have left undone those which I ought to have done. Nowadays that seems to me to be the real definition of a holiday

For once he was not doing very much holiday work, though he was filing at the Rede Lecture to get it into shape for publication. The examinations for the Science and Art Department were over, and indeed he writes to Sir M. Foster :—

Don't bother your head about the balance—now or hereafter. To tell you the truth I do so little in the Examiner business that I am getting ashamed of taking even the retaining fee, and you will do me a favour if you will ease my conscience.

A week of fishery business in South Wales and Devon had "a good deal of holiday in it." For the rest—

I have just been put on Senate of University of London [a Crown nomination]. I tried hard to get Lord Granville to let me off—in fact I told him I could not attend the meetings except now and then, but there was no escape. I must have a talk with you about what is to be done there.

Item.—There is a new Fishery Commission that I also strongly objected to, but had to cave in so far as I agreed to attend some meetings in latter half of September.

On this occasion Lord Granville had written back :—

11 CARLTON HOUSE TERRACE,
July 28, 1883.

MY DEAR PROFESSOR HUXLEY—Clay, the great whist player, once made a mistake and said to his partner, "My brain is softening," the latter answered "Never mind, I will give you £10,000 down for it, just as it is."

On that principle and backed up by Paget I shall write to Harcourt on Monday.—Yours sincerely,

GRANVILLE.

The Commission of course cut short the stay at Milford, and on September 12, he writes :—

We shall leave this on Friday as my wife has some fal-lals to look after before we start for the north on Monday.

The worst of it is that it is not at all certain that the Commission will meet and do any work. However I am pledged to go, and I daresay that Brechin Castle is a very pleasant place to stay in.

Lastly, he was thinking over the obituary notice of Darwin which he had undertaken to write for the Royal Society—though it did not appear till 1888—that on F. Balfour being written by Sir M. Foster.

HIGHCROFT HOUSE, MILFORD, GODALMING,
Aug. 27, 1883.

MY DEAR FOSTER—I do not see anything to add or alter to what you have said about Balfour, except to get rid of that terrible word "urinogenital," which he invented, and I believe I once adopted, out of mere sympathy I suppose.

Darwin is on my mind, and I will see what can be

done here by and by. Up to the present I have been filing away at the Rede Lecture. I believe that getting things into shape takes me more and more trouble as I get older—whether it is a loss of faculty or an increase of fastidiousness I can't say—but at any rate it costs me more time and trouble to get things finished—and when they are done I should prefer burning to publishing them.

Haven't you any suggestions to offer for Anniversary address? I think the Secretaries ought to draw it up, like a Queen's speech.

Mind we have a talk some day about University of London. I suppose you want an English Sorbonne. I have thought of it at times, but the Philistines are strong.

Weather jolly, but altogether too hot for anything but lying on the grass " under the tegmination of the patulous fage," as the poet observes.—Ever yours very faithfully,

T. H. HUXLEY.

The remaining letters of this year are for the most part on Royal Society business, some of which, touching the anniversary dinner, may be quoted :—

4 MARLBOROUGH PLACE, *Nov.* 10, 1883.

MY DEAR FOSTER—. . . I have been trying to get some political and other swells to come to the dinner. Lord Mayor is coming—thought I would ask him on account of City and Guilds business—Lord Chancellor, probably, Courtney, M.P., promised, and I made the greatest blunder I ever made in all my life by thoughtlessly writing to ask Chamberlain (!!!) utterly forgetting the row with Tyndall.[1]

By the mercy of Providence he can't come this year, though I must ask him next (if I am not kicked out for

---

[1] Concerning the Lighthouses.

my sins before that), as he is anxious to come.   Science
ought to be in league with the Radicals. . . .   Ever yours,
T. H. HUXLEY.

He had made prompt confession as soon as he
discovered his mistake, to Tyndall himself, who
ultimately came to the dinner and proposed the
health of his old friend Hirst.

4 MARLBOROUGH PLACE, *Nov.* 9, 1883.

MY DEAR TYNDALL—I have been going to write to you
for two or three days to ask you to propose Hirst's health
as Royal Medallist on the 30th November.   I am sure
your doing so would give an extra value to the medal to
him.

But now I realise the position of those poor devils
I have seen in lunatic asylums and who believed they
have committed the unforgivable sin.   It came upon me
suddenly in Waterloo Place this evening, that I had done
so ; and I went straight to the Royal Institution to make
confession, and if possible get absolution.   But I heard
you had gone to Hindhead, and so I write.

Yesterday I was sending some invitations to the dinner
on the 30th, and thinking to please the Society I made
a shot at some ministers.   The only two I know much
about are Harcourt and Chamberlain, and the devil (in
whom I now firmly believe) put it into my head to write
to both.

The enormous stupidity of which I had been guilty in
asking Chamberlain under the circumstances, and the sort
of construction you and others might put upon it, never
entered my head till this afternoon.   It really made me
ill, and I went straight to find you.   If Providence is good
to me the letter will miscarry and he won't come.   But
anyhow I want you to know that I have been idiotically
stupid, and that I shall wish the Presidency and the

dinner and everything connected with it at the bottom of
the sea, if you are as much disgusted with me as you have
a perfect right to be.—Ever yours faithfully,

T. H. HUXLEY.

The following refers to the Tyneside Sunday
Lecture Society at Newcastle, which had invited
him to become one of its vice-presidents :—

4 MARLBOROUGH PLACE, N.W.,
Dec. 30, 1883.

MY DEAR MORLEY—The Newcastle people wrote to me
some time ago telling me that Sir W. Armstrong was
going to be their President.[1] Armstrong is an old friend
of mine, so I wrote to him to make inquiries. He told me
that he was not going to be President, and knew nothing
about the people who were getting up the Society. So I
declined to have anything to do with it.

However, the case is altered now that you are in the
swim. You have no gods to swear by, unfortunately ;
but if you will affirm, in the name of X, that under no
circumstances shall I be called upon to do anything, they
may have my name among the V.-P.'s and much good may
it do them.

All our good wishes to you and yours. The great
thing one has to wish for as time goes on is vigour as long
as one lives, and death as soon as vigour flags.

It is a curious thing that I find my dislike to the
thought of extinction increasing as I get older and nearer
the goal.

It flashes across me at all sorts of times with a sort of
horror that in 1900 I shall probably know no more of

---

[1] The actual words of the Secretary were "We have asked Sir
W. Armstrong to be President," and Huxley was mistaken in
supposing this intimation to imply that, as generally happens in
such cases, Sir William had previously intimated his willingness to
accept the position if formally asked.

what is going on than I did in 1800. I had sooner be in
hell a good deal—at any rate in one of the upper circles
where the climate and company are not too trying. I
wonder if you are plagued in this way.—Ever yours,

T. H. H

The following letters, to his family or to intimate
friends, are in lighter vein. The first is to Sir M.
Foster; the concluding item of information in reply
to several inquiries. The Royal Society wished some
borings made in Egypt to determine the depth of the
stratum of Nile mud :—

The Egyptian exploration society is wholly archæo-
logical—at least from the cut of it I have no doubt it is
so—and they want all their money to find out the pawn-
brokers' shops which Israel kept in Pithom and Rameses
—and then went off with the pledges.

This is the real reason why Pharaoh and his host pursued them ; and then Moses and Aaron bribed the post-
boys to take out the linch pins.

That is the real story of the Exodus—as detailed in a
recently discovered papyrus which neither Brugsch nor
Maspero have as yet got hold of.

TO HIS YOUNGEST DAUGHTER

4 MARLBOROUGH PLACE, N.W.,
April 12, 1883.

DEAREST PABELUNZA—I was quite overcome to-day to
find that you had vanished without a parting embrace to
your "faded but fascinating"[1] parent. I clean forgot

---

[1] A fragment of feminine conversation overheard at the Dublin
meeting of the British Association, 1878. "Oh, there comes
Professor Huxley : faded, but still fascinating."

you were going to leave this peaceful village for the whirl of Gloucester dissipation this morning—and the traces of weeping on your visage, which should have reminded me of our imminent parting, were absent.

My dear, I should like to have given you some good counsel. You are but a simple village maiden—don't be taken by the appearance of anybody. Consult your father —inclosing photograph and measurement (in inches)—in any case of difficulty.

Also give my love to the matron your sister, and tell her to look sharp after you. Treat her with more respect than you do your venerable P.—whose life will be gloom hidden by a film of heartless jests till you return.

Item.—Kisses to Ria and Co.—Your desolated Pater.

To his Eldest Daughter

4 MARLBOROUGH PLACE, *May* 6, 1883.

DEAREST JESS—Best thanks for your good wishes— considering all things, I am a hale old gentleman. But I had to speak last night at the Academy dinner, and either that or the quantity of cigars I smoked, following the bad example of our friend " Wales," has left me rather shaky to-day. It was trying, because Jack's capital portrait was hanging just behind me—and somebody remarked that it was a better likeness of me than I was. If you begin to think of that it is rather confusing.

I am grieved to have such accounts of Ethel, and have lectured her accordingly. She threatens reprisals on you —and altogether is in a more saucy and irrepressible state than when she left.

M— is still in bed, though better—I am afraid she won't be able to go to Court next week. You see we are getting grand.

I hear great accounts of the children (Ria and Buzzer) and mean to cut out T'other Governor when you bring them up.

As we did not see Fred the other day, the family is inclined to think that the salmon disagreed with him !—
Ever your loving father,   T. H. HUXLEY.

4 MARLBOROUGH PLACE, *May* 10, 1883.

MY DEAR MRS. TYNDALL—If you will give me a bit of mutton at one o'clock I shall be very much your debtor, but as I have business to attend to afterwards at the Home Office I must stipulate that my intellect be not imperilled by those seductive evil genii who are apt to make their appearance at your lunch table.[1]

M. is getting better, but I cannot let her be out at night yet. She thinks she is to be allowed to go to the International Exhibition business on Saturday; but if the temperature does not rise very considerably I shall have two words to say to that.—Ever yours very sincerely,
                         T. H. HUXLEY.

I shall be alone. Do you think that I am "subdued to that I work in," and like an oyster, carry my brood about beneath my mantle ?

---

[1] This is accompanied by a sketch of a champagne bottle in the character of a demon.

CHAPTER XV

1884

FROM this time forward the burden of ill-health grew slowly and steadily. Dyspepsia and the hypochondriacal depression which follows in its train, again attacked Huxley as they had attacked him twelve years before, though this time the physical misery was perhaps less. His energy was sapped; when his official work was over, he could hardly bring himself to renew the investigations in which he had always delighted. To stoop over the microscope was a physical discomfort; he began to devote himself more exclusively to the reading of philosophy and critical theology. This was the time of which Sir M. Foster writes that "there was something working in him which made his hand, when turned to anatomical science, so heavy that he could not lift it. Not even that which was so strong within him, the duty of fulfilling a promise, could bring him to the work."

Up to the beginning of October, he went on with his official work, the lectures at South Kensington, the business as President of the Royal Society, and

ex officio Trustee of the British Museum ; the duties
connected with the Inspectorship of Fisheries, the
City and Guilds Technical Education Committee, and
the University of London, and delivered the opening
address at the London Hospital Medical School, on
"The State and the Medical Profession" (*Coll. Ess.*
iii. 323), his health meanwhile growing less and less
satisfactory.   He dropped minor offices, such as the
Presidency of the National Association of Science
Teachers, which, he considered, needed more careful
supervision than he was able to give, and meditated
retiring from part at least of his main duties, when
he was ordered abroad at a moment's notice for first
one, then another, and yet a third period of two
months.   But he did not definitely retire until this
rest had proved ineffectual to fit him again for active
work

The President of the Royal Society is, as mentioned above, an *ex officio* Trustee of the British
Museum, so that now, as again in 1888, circumstances
at length brought about the state of affairs which
Huxley had once indicated—half jestingly—to
Robert Lowe, who inquired of him what would be
the best course to adopt with respect to the Natural
History collections of the British Museum :—"Make
me a Trustee and Flower director."   At this
moment, the question of an official residence for
the Director of the Natural History Museum was
under discussion with the Treasury, and he
writes :—

Feb. 29, 1884.

MY DEAR FLOWER—I am particularly glad to hear your news. "Ville qui parle et femme qui écoute se rendent," says the wicked proverb—and it is true of Chancellors of the Exchequer.—Ever yours very faithfully,   T. H. HUXLEY.

A pendent to this is a letter of congratulation to Sir Henry Roscoe on his knighthood :—

SCIENCE AND ART DEPARTMENT, S.K.,
July 7, 1884.

MY DEAR ROSCOE—I am very glad to see that the Government has had the grace to make some acknowledgment of their obligation to you, and I wish you and "my lady" long enjoyment of your honours. I don't know if you are gazetted yet, so I don't indicate them outside.—Ever yours very faithfully,   T. H. HUXLEY.

P.S.—I wrote some weeks ago to the Secretary of the National Association of Science Teachers to say that I must give up the Presidency. I had come to the conclusion that the Association wants sharp looking after, and that I can't undertake that business.

P.S. 2.—Shall I tell you what your great affliction henceforward will be ? It will be to hear yourself called Sr'enery Roscoe by the flunkies who announce you.

Her Ladyship will please take note of this crumpled roso loaf—I am sure of its annoying her.

The following letter, with its comparison of life to a whirlpool and its acknowledgment of the widespread tendency in mankind to make idols, was written in answer to some inquiries from Lady Welby :—

April 8, 1884.

Your letter requires consideration, and I have had very little leisure lately. Whether motion disintegrates or integrates is, I apprehend, a question of conditions. A whirlpool in a stream may remain in the same spot for any imaginable time. Yet it is the effect of the motion of the particles of the water in that spot which continually integrate themselves into the whirlpool and disintegrate themselves from it. The whirlpool is permanent while the conditions last, though its constituents incessantly change. Living bodies are just such whirlpools. Matter sets into them in the shape of food,—sets out of them in the shape of waste products. Their individuality lies in the constant maintenance of a characteristic form, not in the preservation of material identity. I do not know anything about "vitality" except as a name for certain phenomena like "electricity" or "gravitation." As you get deeper into scientific questions you will find that "Name ist Schall und Rauch" even more emphatically than Faust says it is in Theology. Most of us are idolators, and ascribe divine powers to the abstractions "Force," "Gravity," "Vitality," which our own brains have created. I do not know anything about "inert" things in nature. If we reduce the world to matter and motion, the matter is not "inert," inasmuch as the same amount of motion affects different kinds of matter in different ways. To go back to my own illustration. The fabric of the watch is not inert, every particle of it is in violent and rapid motion, and the winding-up simply perturbs the whole infinitely complicated system in a particular fashion. Equilibrium means death, because life is a succession of changes, while a changing equilibrium is a contradiction in terms. I am not at all clear that a living being is comparable to a machine running down. On this side of the question the whirlpool affords a better parallel than the watch. If you dam the stream above or below, the whirlpool dies; just as the living being does if you cut off its food, or choke it with its own waste products. And if you alter the sides or bottom of the stream you may kill the whirlpool, just as you kill the animal by interfering with its structure. Heat and oxidation as a source of heat appear to supply energy to the living machine, the molecular structure of the germ furnishing the "sides and bottom of the stream," that is, determining the results which the energy supplied shall produce.

Mr. Ashby writes like a man who knows what he is talking about. His exposition appears to me to be essentially sound and extremely well put. I wish there were more sanitary officers of the same stamp. Mr. Spencer is a very admirable writer, and I set great store by his works. But we are very old friends, and he has endured me as a sort of "devil's-advocate" for thirty-odd years. He thinks that if I can pick no holes in what he says he is safe. But I pick a great many holes, and we agree to differ.

Between April and September, Fishery business took him out of London for no less than forty-three days, first to Cornwall, then in May to Brixham, in June to Cumberland and Yorkshire, in July to Chester, and in September to South Devon, Cornwall, and Wales. A few extracts from his letters home may be given. Just before starting, he writes from Marlborough Place to Rogate, where his wife and one of his daughters were staying :—

April 8.—The weather turned wonderfully muggy here this morning, and turned me into wet paper. But I contrived to make a "neat and appropriate" in presenting old Hird with his testimonial. Fayrer and I were students under him forty years ago, and as we stood together it was a question which was the greyest old chap.

April 14.—I have almost given up reading the Egyptian news, I am so disgusted with the whole business. I saw several pieces of land to let for building purposes about Falmouth, but did not buy. (This was to twit his wife with her constant desire that he should buy a bit of land in the country to settle upon in their old age.)

April 18.—You don't say when you go back, so I direct this to Rogate. I shall expect to see you quite set up. We must begin to think seriously about getting out of the hurly-burly a year or two hence, and having an Indian summer together in peace and quietness.

April 15, *Sunday, Falmouth.*—I went out at ten o'clock this morning, and did not get back till near seven. But I got a cup of tea and some bread and butter in a country village, and by the help of that and many pipes supported nature. There was a bitter east wind blowing, but the day was lovely otherwise, and by judicious dodging in coves and creeks and sandy bays, I escaped the wind and absorbed a prodigious quantity of sunshine.

I took a volume of the *Decline and Fall of the Roman Empire* with me. I had not read the famous 15th and 16th chapters for ages, and I lay on the sands and enjoyed them properly. A lady came and spoke to me as I returned, who knew L. at Oxford very well—can't recollect her name—and her father and mother are here, and I have just been spending an hour with them. Also a man who sat by me at dinner knew me from Jack's portrait. So my incognito is not very good. I feel quite set up by my day's wanderings.

May 11, *Torquay.*—We went over to Brixham yesterday to hold an inquiry, getting back here to an eight-o'clock or nearer nine dinner. . . . Dalhousie has discovered that the officer now in command of the *Britannia*

is somebody whom he does *not* know, so we gave up
going to Dartmouth and agreed to have a lazy day here.
It is the most exquisite summer weather you can
imagine, and I have been basking in the sun all the
morning and dreamily looking over the view of the
lovely bay which is looking its best—but take it all
round it does not come up to Lynton. Dalhousie is more
likeable than ever, and I am just going out for a stroll
with him.

June 24.—I left Keswick this morning for Cocker-
mouth, took the chair at my meeting punctually at
twelve, sat six mortal hours listening to evidence, nine-
tenths of which was superfluous—and turning my lawyer
faculty to account in sifting the grains of fact out of the
other tenth.

June 25, *Leeds.*—. . . We had a long drive to a
village called Harewood on the Wharfe. There is a big
Lord lives there—Earl of Harewood—and he and his
ancestors must have taken great care of their tenants, for
the labourers' houses are the best I ever saw. . . . I cut
out the enclosed from the *Standard* the other day to
amuse you, but have forgotten to send it before.[1] I
think we will be "Markishes," the lower grades are
getting common.

June 27.—. . . I had a long day's inspection of the
Wharfe yesterday, attended a meeting of the landed
proprietors at Ottley to tell them what they must do if
they would get salmon up their river. . . .

I shall leave here to-morrow morning, go on to
Skipton, whence seven or eight miles' drive will take me
to Linton where there is an obstruction in the river I
want to see. In the afternoon I shall come home from
Skipton, but I don't know exactly by what train. As
far as I see, I ought to be home by about 10.30, and you

---

[1] Apparently announcing that he was about to accept a title.
I have not been able to trace the paragraph.

may have something light for supper, as the "course of true feeding is not likely to run smooth"—to-morrow.

In August he went again to the corner of Surrey which he had enjoyed so much the year before. Here, in the intervals of suffering under the hands of the dentist, he worked at preparing a new edition of the *Elementary Physiology* with Sir M. Foster, alternating with fresh studies in critical theology.

The following letters reflect his occupations at this time, together with his desire, strongly combated by his friend, of resigning the Presidency of the Royal Society immediately.

> HIGHCROFT HOUSE, MILFORD,
> GODALMING, *Aug.* 9, 1884.
>
> MY DEAR FOSTER—I had to go up to town on Friday, and yesterday I went and had all my remaining teeth out, and came down here again with a shrewd suspicion that I was really drunk and incapable, however respectable I might look outwardly. At present I can't eat at all, and *I can't smoke with any comfort.* For once I don't mind using italics.
>
> Item.—I send the two cuts.
>
> Heaven be praised! I had brought down no copy of Physiology with me, so could not attend to your proof. Got it yesterday, so I am now at your mercy.
>
> But I have gone over the proofs now, and send you a deuce of a lot of suggestions.
>
> Just think over additions to smell and taste to bring these into harmony.
>
> The Saints salute you. I am principally occupied in studying the Gospels.—Ever yours,
>
> T. H. HUXLEY.

HIGHCROFT HOUSE, MILFORD,
GODALMING, *Aug.* 26, 1884.

DEARLY BELOVED—I have been going over the ear chapter this morning, and, as you will see, have suggested some additions. Those about the lamina spiralis are certainly necessary—*illus.* substitution of trihedral for triangular.[1] I want also very much to get into heads of students that in sensation it is all modes of motion up to and in sensorium, and that the generation of feeling is the specific reaction of a particle of the sensorium when stimulated, just as contraction, etc., is the specific reaction of a muscular fibre when stimulated by its nerve. The psychologists make the fools of themselves they do because they have never mastered this elementary fact. But I am not sure whether I have put it well, and I wish you would give your mind to it. As for me I have not had much mind to give lately—a fortnight's spoon-meat reduced me to inanity, and I am only just picking up again. However, I walked ten miles yesterday afternoon, so there is not much the matter.

I will see what I can do with the histology business.[2] I wanted to re-write it, but I am not sure yet whether I shall be able.

Between ourselves, I have pretty well made up my mind to clear out of everything next year, R.S. included. I loathe the thought of wasting any more of my life in endless distractions—and so long as I live in London there is no escape for me. I have half a mind to live abroad for six months in the year.—Ever yours,

T. H. H.

---

[1] On Sept. 8, he writes :—"I have been laughing over my 'trihedron.' It is a regular bull."

[2] "Most of our examinees" (he writes on Sept. 5) "have not a notion of what histology means at present. I think it will be good for other folks to get it into their heads that it is not all sections and carmine."

I enclose letter from Deutsch lunatic to go before
Council and be answered by Foreign Secretary.

HIGHCROFT HOUSE, MILFORD,
GODALMING, *Aug.* 29, 1884.

DEARLY BELOVED—I enclose the proofs, having
mustered up volition enough to go over them at once. I
think the alterations will be great improvements. I see
you interpret yourself about the movements of the
larynx.

As to the histology, I shall have a shot at it, but if I
do not send you MS. in a week's time, go ahead. I am
perplexed about the illustrations, but I see nothing for it
but to have new ones in all the cases which you have
marked. Have you anybody in Cambridge who can
draw the things from preparations?

You are like Trochu with your "plan," and I am
anxious to learn it. But have you reflected, 1st, that I
am getting deafer and deafer, and that I cannot hear
what is said at the council table and in the Society's
rooms half the time people are speaking? and 2nd, that
so long as I am President, so long must I be at the beck
and call of everything that turns up in relation to the
interests of science. So long as I am in the chair, I cannot be a *fainéant* or refuse to do anything and everything incidental to the position.

My notion is to get away for six months, so as to
break with the "world, the flesh, and the devil" of
London, for all which I have conceived a perfect loath-
ing. Six months is long enough for anybody to be
forgotten twice over by everybody but personal friends.

I am contemplating a winter in Italy, but I shall
keep on my house for Harry's sake and as a *pied à terre*
in London, and in the summer come and look at you at
Burlington House, as the old soap-boiler used to visit the
factory. I shall feel like the man out of whom the
legion of devils departed when he looked at the gambades

of the two thousand pigs going at express speed for the waters of Tiberias.

By the way, did you ever read that preposterous and immoral story carefully ? It is one of the best attested of the miracles.

When I have retired from the chair (which I must not scandalise) I shall write a lay sermon on the text. It will be impressive.

My wife sends her love, and says she has her eye on you. She is all for retirement.—Ever yours,

I am very sorry to hear of poor Mangles' death, but I suppose there was no other chance.　　　T. H. H.

In September he hails with delight some intermission of the constant depression under which he has been labouring, and writes :—

So long as I sit still and write or read I am all right, otherwise not good for much, which is odd, considering that I eat, drink, and sleep like a top. I suppose that everybody starts with a certain capital of life-stuff, and that expensive habits have reduced mine.

And again :—

I have been very shaky for the last few weeks, but I am picking up again, and hope to come up smiling for the winter's punishment.

There was nothing to drink last night, so I had some tea ! with my dinner—smoked a pipe or two—slept better than usual, and woke without blue devils for the first time for a week ! ! ! Query, is that the effect of tea or baccy ? I shall try them again. We are fearfully and wonderfully made, especially in the stomach—which is altogether past finding out.

Still, his humour would flash out in the midst of

his troubles; he writes in answer to a string of semi-official inquiries from Sir J. Donnelly:—

HIGHCROFT HOUSE, MILFORD,
GODALMING.

SIR—In reply to your letter of the 9th Aug. (666), I have the honour to state—

1. That I am here.
2. That I have (a) had all my teeth out; (b) partially sprained my right thumb; (c) am very hot; (d) can't smoke with comfort; whence I may leave even official intelligence to construct an answer to your second inquiry.
3. Your third question is already answered under 2a. Not writing might be accounted for by 2b, but unfortunately the sprain is not bad enough—and "laziness, sheer laziness" is the proper answer.

I am prepared to take a solemn affidavit that I told you and Macgregor where I was coming many times, and moreover that I distinctly formed the intention of leaving my address in writing—according to those official instructions which I always fulfil.

If the intention was not carried out, its blood be upon its own head—I wash my hands of it, as Pilate did.

4. As to the question whether I *want* my letters I can sincerely declare that I don't—would in fact much rather not see them. But I suppose for all that they had better be sent.
5. I hope Macgregor's question is not a hard one—spoon-meat does not carry you beyond words of one syllable.

On Friday I signalised my last dinner for the next three weeks by going to meet the G.O.M. I sat next him, and he was as lively as a bird.

Very sorry to hear about your house. You will have to set up a van with a brass knocker and anchor on our common.—Ever yours, T. H. HUXLEY.

By the beginning of September he had made up his mind that he ought before long to retire from active life. The first person to be told of his resolution was the head of the Science and Art Department, with whom he had worked so long at South Kensington.

HIGHCROFT HOUSE, MILFORD, GODALMING,
Sept. 3, 1884.

MY DEAR DONNELLY—I was very glad to have news of you yesterday. I gather you are thriving, notwithstanding the appalling title of your place of refuge. I should have preferred " blow the cold " to " Cold blow "— but there is no accounting for tastes.

I have been going and going to write to you for a week past to tell you of a notion that has been maturing in my mind for some time, and that I ought to let you know of before anybody else. I find myself distinctly aged—tired out body and soul, and for the first time in my life fairly afraid of the work that lies before me in the next nine months. Physically, I have nothing much. to complain of except weariness—and for purely mental work, I think I am good for something yet. I am morally and mentally sick of society and societies— committees, councils—bother about details and general worry and waste of time.

I feel as if more than another year of it would be the death of me. Next May I shall be sixty, and have been thirty-one mortal years in my present office in the School. Surely I may sing my *nunc dimittis* with a good conscience. I am strongly inclined to announce to the Royal Society in November that the chair will be vacant that day twelvemonth—to resign my Government posts at midsummer, and go away and spend the winter in Italy—so that I may be out of reach of all the turmoil of London. The only thing I don't like is the notion of leaving

you without such support as I can give in the School.
No one knows better than I do how completely it is your
work and how gallantly you have borne the trouble and
responsibility connected with it. But what am I to do?
I must give up all or nothing—and I shall certainly come
to grief if I do not have a long rest.

Pray tell me what you think about it all.

My wife has written to Mrs. Donnelly and told her
the news.—Ever yours very faithfully,

T. H. HUXLEY.

Read Hobbes if you want to get hard sense in good
English.

HIGHCROFT HOUSE, MILFORD, GODALMING,
Sept. 10, 1884.

MY DEAR DONNELLY—Many thanks for your kind
letter. I feel rather like a deserter, and am glad of any
crumbs of comfort.

Cartwright has done wonders for me, and I can already
eat most things (I draw the line at tough crusts). I have
not even my old enemy, dyspepsia—but eat, drink, and
sleep like a top.

And withal I am as tired as if I were hard at work,
and shirk walking.

So far as I can make out there is not the slightest sign
of organic disease anywhere, but I will get Clark to over-
haul me when I go back to town. Sometimes I am
inclined to suspect that it is all sham and laziness—but
then why the deuce should I want to sham and be lazy.

Somebody started a charming theory years ago—that
as you get older and lose volition, primitive evil tendencies,
heretofore mastered, come out and show themselves. A
nice prospect for venerable old gentlemen !

Perhaps my crust of industry is denuded, and the
primitive rock of sloth is cropping out.

But enough of this egotistical invalidism.

How wonderfully Gordon is holding his own. I should like to see him lick the Mahdi into fits before Wolseley gets up. You despise the Jews, but Gordon is more like one of the Maccabees or Bar-Kochba than any sort of modern man.

My wife sends love to both of you, and says you are (in feminine language) " a dear thing in friends."—Ever yours very faithfully,   T. H. HUXLEY.

HOME OFFICE, *Sept.* 18, 1884.

MY DEAR DONNELLY—We have struck our camp at Milford, and I am going down to Devonshire and Cornwall to-morrow—partly on Fishery business, partly to see if I can shake myself straighter by change of air. I am possessed by seven devils—not only blue, but of the deepest indigo—and I shall try to transplant them into a herd of Cornish swine.

The only thing that comforts me is Gordon's telegrams. Did ever a poor devil of a Government have such a subordinate before ? He is the most refreshing personality of this generation.

I shall be back by 30th September—and I hope in better condition for harness than now.—Ever yours very faithfully,   T. H. HUXLEY.

Replying to General Donnelly's arguments against his resigning all his official posts, he writes :—

DARTMOUTH, *Sept.* 21, 1884.

MY DEAR DONNELLY—Your letters, having made a journey to Penzance (where I told my wife I should go last Friday, but did not, and brought up here instead) turned up this morning.

I am glad to have seen Lord Carlingford's letter, and I am very much obliged to him for his kind expressions. Assuredly I will not decide hastily.

Now for your letter—I am all for letters in these

matters. Not that we are either of us "impatient and irritable listeners"—oh dear, no! "I have my faults," as the miser said, "but *avarice* is not one of them"—and we have our faults too, but notoriously they lie in the direction of long-suffering and apathy.

Nevertheless there is a good deal to be said for writing. *Mine* is· itself a discipline in patience for my correspondent.

Imprimis. I scorn all your chaff about Society. My great object for years has been to keep out of it, not to go into it. Just you wait till the Misses Donnelly grow up —I trust there may be five or ten of them—and see what will happen to you. But apart from this, so long as I live in London, so long will it be practically impossible for me to keep out of dining and giving of dinners—and you know that just as well as I do.

2nd. I mean to give up the Presidency, but don't see my way to doing so next St. Andrew's Day. I wish I could—but I must deal fairly by the Society.

3rd. The suggestion of the holiday at Christmas is the most sensible thing you have said. I could get six weeks under the new arrangement (*Botany*, January and half February) without interfering with my lectures at all. But then there is the blessed Home Office to consider. There might be civil war between the net men and the rod men in six weeks, all over the country, without my mild influence.

4th. I must give up my Inspectorship. The mere thought of having to occupy myself with the squabbles of these idiots of country squireens and poachers makes me sick—and is, I believe, the chief cause of the morbid state of my mucous membranes.

All this week shall I be occupied in hearing one Jackass contradict another Jackass about questions which are of no importance.

I would almost as soon be in the House of Commons.

Now see how reasonable I am. I agree with you (a)

that I must get out of the hurly-burly of society; (*b*)
that I must get out of the Presidency; (*c*) that I must
get out of the Inspectorship, or rather I agree with
myself on that matter, you having expressed no opinion.

That being so, it seems to me that I must, willy-nilly,
give up S.K. For—and here is the point you had in
your mind when you lamented your possible impatience
about something I might say—I swear by all the gods
that are not mine, nothing shall induce me to apply to
the Treasury for anything but the pound of flesh to which
I am entitled.

Nothing ever disgusted me more than being the subject
of a battle with the Treasury over the H.O. appointment
—which I should have thrown up if I could have done
so with decency to Harcourt.

It's just as well for me I couldn't, but it left a nasty
taste.

I don't want to leave the School, and should be very
glad to remain as Dean, for many reasons. But what I
don't see is how I am to do that and make my escape
from the thousand and one entanglements—which seem
to me to come upon me quite irrespectively of any office
I hold—or how I am to go on living in London as a
(financially) decayed philosopher.

I really see nothing for it but to take my pension and
go and spend the winter of 1885-86 in Italy. I hear
one can be a regular swell there on £1000 a year.

Six months' absence is oblivion, and I shall take to a
new line of work, and one which will greatly meet your
approval.

As to X—— I am not a-going to—not being given
to hopeless enterprises. That rough customer at Dublin
is the only man who occurs to me. I can't think of his
name, but that is part of my general unfitness.

. . . . I suppose I shall chaff somebody on my death-bed.
But I am out of heart to think of the end of the lunches
in the sacred corner.—Ever yours, T. H. HUXLEY.

On the 21st he writes home about the steps he had begun to take with respect to giving up part of his official work.

> I have had a long letter from Donnelly. He had told Lord Carlingford of my plans, and incloses a letter from Lord Carlingford to him, trusting I will not hastily decide, and with some pretty phrases about "support and honour" I give to the School. Donnelly is very anxious I should hold on to the School, if only as Dean, and wants me in any case to take two months' holiday at Christmas. Of course he looks on the R.S. as the root of all evil. Foster *per contra* looks on the School as the deuce, but would have me stick by the Royal Society like grim death.
>
> The only moral obligation that weighs with me is that which I feel under, to deal fairly by Donnelly and the School. You must not argue against this, as rightly or wrongly I am certain that if I deserted the School hastily, or if I did not do all that I can to requite Donnelly for the plucky way in which he has stood by it and me for the last dozen years, I should never shake off the feeling that I had behaved badly. And as I am much given to brooding over my misdeeds, I don't want you to increase the number of my hell-hounds. You must help me in this . . and if I am Quixotic, play Sancho for the nonce.

CHAPTER XVI

1884-85

TOWARDS the end of September he went to the West country to try to improve his health before the session began again in London. Thus he writes, on Sept. 26, to Mr. W. F. Collier, who had invited him to Horrabridge, and on the 27th to Sir M. Foster :—

FOWEY, *Sept.* 26, 1884.

Many thanks for the kind offer in your letter, which has followed me here. But I have not been on the track you might naturally have supposed I had followed. I have been trying to combine hygiene with business, and betook myself, in the first place, to Dartmouth, afterwards to Totnes, and then came on here. From this base of operations I could easily reach all my places of meeting. To-morrow I have to go to Bodmin, but I shall return here, and if the weather is fine (raining cats and dogs at present), I may remain a day or two to take in stock of fresh air before commencing the London campaign.

I am very glad to hear that your health has improved so much. You must feel quite proud to be such an interesting "case." If I set a good example myself I would venture to warn you against spending five shillings' worth of strength on the ground of improvement to the extent of half-a-crown.

I am not quite clear as to the extent to which my children have colonised Woodtown at present. But it seems to me that there must be three or four Huxleys (free or in combination, as the chemists say) about the premises. Please give them the paternal benediction; and with very kind remembrances to Mrs. Collier, believe me —Yours very faithfully,   T. H. HUXLEY.

FOWEY HOTEL, FOWEY, CORNWALL,
Sept. 27, 1884.

MY DEAR FOSTER—I return your proof, with a few trifling suggestions here and there. . . .

I fancy we may regard the award as practically settled, and a very good award it will be.

The address is beginning to loom in the distance. I have half a mind to devote some part of it to a sketch of the recent novelties in histology touching the nucleus question and molecular physiology.

My wife sent me your letter. By all means let us have a confabulation as soon as I get back and settle what is to be done with the "aged P."

I am not sure that I shall be at home before the end of the week. My lectures do not begin till next week, and the faithful Howes can start the practical work without me, so that if I find myself picking up any good in these parts, I shall probably linger here or hereabouts. But a good deal will depend on the weather—inside as well as outside. I am convinced that the prophet Jeremiah (whose works I have been studying) must have been a flatulent dyspeptic—there is so much agreement between his views and mine.—Ever yours,

T. H. HUXLEY.

But the net result of this holiday is summed up in a note, of October 5, to Sir M. Foster :—

I got better while I was in Cornwall and Wales, and, at present, I don't think there is anything the matter

with me except a profound disinclination to work. I
never before knew the proper sense of the term "vis
inertiæ."

And writing in the same strain to Sir J. Evans,
he adds :—

But I have a notion that if I do not take a long spell
of absolute rest before long I shall come to grief. However, getting into harness again may prove a tonic—it
often does, *e.g.* in the case of cab-horses.

Three days later he found himself ordered to leave
England immediately, under pain of a hopeless breakdown.

4 MARLBOROUGH PLACE,
Oct. 8, 1884.

MY DEAR FOSTER—We shall be very glad to see you
on Friday. I came to the conclusion that I had better
put myself in Clark's hands again, and he has been here
this evening overhauling me for an hour.

He says there is nothing wrong except a slight affection of the liver and general nervous depression, but
that if I go on the latter will get steadily worse and
become troublesome. He insists on my going away to
the South and doing nothing but amuse myself for three
or four months.

This is the devil to pay, but I cannot honestly say
that I think he is wrong. Moreover, I promised the
wife to abide by his decision.

We will talk over what is to be done.—Ever yours,

T. H. HUXLEY.

ATHENÆUM CLUB,
Oct. 13, 1884.

MY DEAR MORLEY—I heartily wish I could be with
you on the 25th, but it is *aliter visum* to somebody,
whether Dis or Diabolis, I can't say.

The fact is, the day after I saw you I had to put myself in Clark's hands, and he ordered me to knock off work and go and amuse myself for three or four months, under penalties of an unpleasant kind.

So I am off to Venice next Wednesday. It is the only tolerably warm place accessible to any one whose wife will not let him go within reach of cholera just at present.

If I am a good boy I am to come back all sound, as there is nothing organic the matter; but I have had enough of the world, the flesh, and the devil, and shall extricate myself from that Trinity as soon as may be. Perhaps I may get within measurable distance of Berkeley (*English Men of Letters*, ed. J.M.) before I die!—Ever yours very faithfully, T. H. HUXLEY.

ATHENÆUM CLUB,
October 18, 1884.

MY DEAR FOSTER—Best thanks for your letter and route. I am giving you a frightful quantity of trouble; but as the old woman (Irish) said to my wife, when she gave her a pair of my old trousers for her husband, "I hope it may be made up to ye in a better world."

She is clear, and I am clear, that there is no reason on my part for not holding on if the Society really wishes I should. But, of course, I must make it easy for the Council to get rid of a *fainéant* President, if they prefer that course.

I wrote to Evans an unofficial letter two days ago, and have had a very kind, straightforward letter from him. He is quite against my resignation. I shall see him this afternoon here. I had to go to my office (Fishery).

Clark's course of physic is lightening my abdominal troubles, but I am preposterously weak with a kind of shabby broken-down indifference to everything.—Ever yours, T. H. H.

The "Indian summer"[1] to which he looked
forward was not to be reached without passing
through a season of more than equinoctial storms
and tempests. His career had reached its highest
point only to be threatened with a speedy close.
He himself did not expect more than two or three
years' longer lease of life, and went by easy stages
to Venice, where he spent eight days. "No place,"
he writes, "could be better fitted for a poor devil as
sick in body and mind as I was when I got there."

Venice itself (he writes to Dr. Foster) just suited me.
I chartered a capital gondolier, and spent most of my
time exploring the Lagoons. Especially I paid a daily
visit to the Lido, and filled my lungs with the sea air,
and rejoiced in the absence of stinks. For Venice is like
her population (at least the male part of it), handsome
but odorous. Did you notice how handsome the young
men are and how little beauty there is among the
women?

I stayed eight days in Venice and then returned by
easy stages first to Padua, where I wanted to see Giotto's
work, then to Verona, and then here (Lugano). Verona
delighted me more than anything I have seen, and we
will spend two other days there as we go back.

As for myself, I really have no positive complaint
now. I eat well and I sleep well, and I should begin
to think I was malingering, if it were not for a sort of
weariness and deadness that hangs about me, accompanied
by a curious nervous irritability.

I expect that this is the upshot of the terrible anxiety
I have had about my daughter M——.

I would give a great deal to be able to escape facing

---

[1] See page 360.

the wedding, for my nervous system is in the condition
of that of a frog under opium.

But my R. must not go off without the paternal
benediction.

For the first three weeks he was alone, his wife
staying to make preparations for the third daughter's
wedding on November 6th, for which occasion he
was to return, afterwards taking her abroad with
him. Unfortunately, just as he started, news was
brought him at the railway station that his second
daughter, whose brilliant gifts and happy marriage
seemed to promise everything for her future, had
been stricken by the beginnings of an insidious and,
as he too truly feared, hopeless disease. Nothing
could have more retarded his own recovery. It was
a bitter grief, referred to only in his most intimate
letters, and, indeed, for a time kept secret even from
the other members of the family. Nothing was to
throw a shade over the brightness of the approaching
wedding.

But on his way home, he writes of that journey :—

I had to bear my incubus, not knowing what might
come next, until I reached Luzern, when I telegraphed
for intelligence, and had my mind set at ease as to the
measures which were being adopted.

I am a tough subject, and have learned to bear a good
deal without crying out; but those four-and-twenty
hours between London and Luzern have taught me that
I have yet a good deal to learn in the way of "grinning
and bearing."

And although he writes, "I would give a good deal not to face a lot of people next week," . . . "I have the feelings of a wounded wild beast and hate the sight of all but my best friends," he hid away his feelings, and made this the occasion for a very witty speech, of which, alas! I remember nothing but a delightfully mixed polyglot exordium in French, German, and Italian, the result, he declared, of his recent excursion to foreign parts, which had obliterated the recollection of his native speech.

During his second absence he appointed his youngest daughter secretary to look after necessary correspondence, about which he forwarded instructions from time to time.

The chief matters of interest in the letters of this period are accounts of health and travel, sometimes serious, more often jesting, for the letters were generally written in the bright intervals between his dark days : business of the Royal Society, and the publication of the new edition of the *Lessons in Elementary Physiology*, upon which he and Dr. Foster had been at work during the autumn. But the four months abroad were not productive of very great good; the weather was unpropitious for an invalid— "as usual, a quite unusual season"—while his mind was oppressed by the reports of his daughter's illness. Under these circumstances recovery was slow and travel comfortless; all the Englishman's love of home breaks out in his letter of April 8, when he set foot again on English soil.

HOTEL DE LONDRES, VERONA,
Nov. 18, 1884.

DEAREST BABS— 1. Why, indeed, do they ask for more? Wait till they send a letter of explanation, and then say that I am out of the country and not expected back for several years.

2. I wholly decline to send in any name to Athenæum. But don't mention it.

3. Society of Arts be bothered, also——

4. Write to Science and Art Club to engage three of the prettiest girls as partners for the evening. They will look very nice as wallflowers.

5. Penny dinners? declined with thanks.

6. Ask the meeting of Herts N.H. Society to come here after next Thursday, when we shall be in Bologna.

Business first, my sweet girl secretary with the curly front; and now for private affairs, though as your mother is covering reams with them, I can only mention a few of the more important which she will forget.

The first is that she has a habit of hiding my shirts so that I am unable to find them when we go away, and the chambermaid comes rushing after us with the garment shamefully displayed.

The second is that she will cover all the room with her things, and I am obliged to establish a military frontier on the table.

The third is that she insists on my buying an Italian cloak. So you will see your venerable pater equipped in this wise.[1] Except in these two particulars, she behaves fairly well to me.

In point of climate, so far, Italy has turned out a fraud. We dare not face Venice, and Mr. Fenili will weep over my defection; but that is better than that we should cough over his satisfaction.

---

[1] Sketch of a cloaked figure like a brigand of melodrama.

I am quite pleased to hear of the theological turn of the family. It must be a drop of blood from one of your eight great-grandfathers, for none of your ancestors that I have known would have developed in this way.

. . . Best love to Nettie and Harry. Tell the former that cabbages do *not* cost 5s. apiece, and the latter that 11 P.M. is the *clôture.*—Ever your affectionate

PATER.

HOTEL BRITANNIQUE, NAPLES,
Nov. 30, 1884.

MY DEAR FOSTER—Which being St. Andrew's Day, I think the expatriated P. ought to give you some account of himself.

We had a prosperous journey to Locarno, but there plumped into bitter cold weather, and got chilled to the bone as the only guests in the big hotel, though they did their best to make us comfortable. I made a shot at bronchitis, but happily failed, and got all right again.

Pallanza was as bad. At Milan temperature at noon 39° F., freezing at night. Verona much the same. Under these circumstances, we concluded to give up Venice and made for Bologna. There found it rather colder. Next Ravenna, where it snowed. However, we made ourselves comfortable in the queer hotel, and rejoiced in the mosaics of that sepulchral marsh.

At Bologna I had assurances that the Sicilian quarantine was going to be taken off at once, and as the reports of the railway travelling and hotels in Calabria were not encouraging, I determined to make for Naples, or rather, by way of extra caution, for Castellamare. All the way to Ancona the Apennines were covered with snow, and much of the plain also. Twenty miles north of Ancona, however, the weather changed to warm summer, and we rejoiced accordingly. At Foggia I found that the one decent hotel that used to exist was non-extant, so we went on to Naples.

Arriving at 10.30 very tired, got humbugged by a
lying Neapolitan, who palmed himself off as the com-
missaire of the Hotel Bristol, and took us into an
omnibus belonging to another hotel, that of the Bristol
being, as he said, "broke." After a drive of three miles
or so got to the Bristol and found it shut up ! After a
series of adventures and a good deal of strong language
on my part, knocked up the people here, who took us in,
though the hotel was in reality shut up like most of
those in Naples.[1]

As usual the weather is "unusual"—hot in the sun,
cold round the corner and at night. Moreover, I found
by yesterday's paper that the beastly Sicilians won't give
up their ten days' quarantine. So all chance of getting
to Catania or Palermo is gone. I am not sure whether
we shall stay here for some time or go to Rome, but at
any rate we shall be here a week.

Dohrn is away getting subsidies in Germany for his
new ship. We inspected the Aquarium this morning.
Eisig and Mayer are in charge. Madame is a good deal
altered in the course of the twelve years that have elapsed
since I saw her, but says she is much better than she was.

As for myself, I got very much better when in North
Italy in spite of the piercing cold. But the fatigue of
the journey from Ancona here, and the worry at the end
of it, did me no good, and I have been seedy for a day
or two. However, I am picking up.

I see one has to be very careful here. We had a
lovely drive yesterday out Pausilippo, but the wife got
chilled and was shaky this morning. However, we got
very good news of our daughter this evening, and that
has set us both up.

My blessing for to-morrow will reach you after date.
Let us hear how everything went off.

Your return in May project is really impracticable on

---

[1] Owing to the cholera and consequent dearth of travellers.

account of the Fishery Report. I cannot be so long
absent from the Home Office whatever I might manage
with S.K.

With our love to Mrs. Foster and you—Ever yours
very faithfully,   T. H. HUXLEY.

This letter, as he says a week later, was written
when he "was rather down in the mouth from the
wretched cold weather, and the wife being laid up
with a bad cold," besides his own ailments.

I find I have to be very careful about night air, but
nothing does me so much good as six or seven miles'
walk between breakfast and lunch—at a good sharp pace.
So I conclude that there cannot be much the matter, and
yet I am always on the edge, so to speak, of that infernal
hypochondria.

We have settled down here very comfortably, and I
do not think we shall care to go any further south.
Madame Dohrn and all the people at the stazione are
very kind, and want to do all sorts of things for us.
The other day we went in the launch to Capri, intending
next day to go to Amalfi. But it threatened bad weather,
so we returned in the evening. The journey knocked
us both up, and we had to get out of another projected
excursion to Ischia to-day. The fact is, I get infinitely
tired with talking to people and can't stand any deviation
from regular and extremely lazy habits. Fancy my
being always in bed by ten o'clock and breakfasting at
nine !

On the 10th, writing to Sir John Evans, who as
Vice-President, was acting in his stead at the Royal
Society, he says :—

In spite of snow on the ground we had three or four
days at Ravenna —which is the most interesting deadly

lively sepulchre of a place I was ever in in my life.
The evolution of modern from ancient art is all there in
a nutshell. . . .

I lead an altogether animal life, except that I have
renewed my old love for Italian. At present I am
rejoicing in the Autobiography of that delightful sinner,
Benvenuto Cellini. I have some notion that there is
such a thing as science somewhere. In fact I am fitting
myself for Neapolitan nobility.

To his Youngest Daughter

HOTEL BRITANNIQUE, NAPLES,
Dec. 22, 1884.

But we have had no letters from home for a week. . . .
Moreover, if we don't hear to-day or to-morrow we shall
begin to speculate on the probability of an earthquake
having swallowed up 4 M. P. "with all the young
barbarians at play—And I their sire trying to get a
Roman holiday" (Byron). For we are going to Rome
to-morrow, having had enough of Naples, the general
effect of which city is such as would be produced by the
sight of a beautiful woman who had not washed or
dressed her hair for a month. Climate, on the whole,
more variable than that of London.

We had a lovely drive three days ago to Cumae, a
perfect summer's day; since then sunshine, heat, cold
wind, calms all durcheinander, with thunder and lightning
last night to complete the variety.

The thermometer and barometer are not fixed to the
walls here, as they would be jerked off by the sudden
changes. At first, it is odd to see them dancing about
the hall. But you soon get used to it, and the porter
sees that they don't break themselves.

With love to Nettie and Harry, and hopes that the
pudding will be good—Ever your loving father,

T. H. HUXLEY.

In January 1885 he went to Rome, whence he writes :—

HOTEL VICTORIA, VIA DEI DUE MACELLI,
ROME, *Jan.* 8, 1885.

MY DEAR FOSTER—We have been here a fortnight very well lodged—south aspect, fireplace, and all the rest of the essentials except sunshine. Of this last there is not much more than in England, and the grey skies day after day are worthy of our native land. Sometimes it rains cats and dogs all day by way of a change—as on Christmas Day—but it is not cold. " Quite exceptional weather," they tell us, but that seems to be the rule everywhere. We have done a respectable amount of gallery-slaving, and I have been amusing myself by picking up the topography of ancient Rome. I was going to say Pagan Rome, but the inappropriateness of the distinction strikes me, papal Rome being much more stupidly and childishly pagan than imperial. I never saw a sadder sight than the kissing a wretched bedizened doll of a Bambino that went on in the Ara Coeli on Twelfth Day. Your Puritan soul would have longed to arise and slay. . . .

As to myself, though it is a very unsatisfactory subject and one I am very tired of bothering my friends about, I am like the farmer at the rent-dinner, and don't find myself much "forrarder." That is to say, I am well for a few days and then all adrift, and have to put myself right by dosing with Clark's pills, which are really invaluable. They will make me believe in those pills I saw advertised in my youth, and which among other things were warranted to cure "the indecision of juries." I really can't make out my own condition. I walked seven or eight miles this morning over Monte Mario and out on the Campagna without any particular fatigue, and yesterday I was as miserable as an owl in sunshine. Something perhaps must be put down to the relapse

which our poor girl had a week ago, and which became
known to us in a terrible way. She had apparently
quite recovered, and arrangements were made for their
going abroad, and now everything is upset. I warned
her husband that this was very likely, but did not sufficiently take the warning to myself.

You are taking a world of trouble for me, and Donnelly
writes I am to do as I like so far as they are concerned.
I have heard nothing from the Home Office, and I
suppose it would be proper for me to write if I want any
more leave. I really hardly know what to do. I can't
say I feel very fit for the hurly-burly of London just now,
but I am not sure that the wholesomest thing for me
would not be at all costs to get back to some engrossing
work. If my poor girl were well, I could perhaps make
something of the *dolce far niente*, but at present one's
mind runs to her when it is not busy in something else.

I expect we shall be here a week or ten days more—
at any rate, this address is safe—afterwards to Florence.

What am I to do in the Riviera? Here and at
Florence there is always some distraction. You see the
problem is complex.

My wife, who is very lively, thanks you for your
letter (which I have answered) and joins with me in love
to Mrs. Foster and yourself.—Ever yours, T. H. H.

Writing on the same day to Sir J. Evans, he
proposed a considerable alteration in the duties of
the Assistant Secretary of the Royal Society.

You know that I served a seven years' apprenticeship
as Secretary, and that experience gave me very solid
grounds for the conviction that, with the present arrangements, a great deal of the time of the Secretaries is wasted
over the almost mechanical drudgery of proof-reading.

He suggests new arrangements, and proceeds :—

At the same time it would be very important to adopt some arrangement by which the *Transactions* papers can be printed independently of one another.

Why should not the papers be paged independently and be numbered for each year. Thus—" Huxley. Idleness and Incapacity in Italy. *Phil. Trans.* 1885. VI."

People grumble at the delay in publication, and are quite right in doing so, though it is impossible under the present system to be more expeditious, and it is not every senior secretary who would slave at the work as Stokes does. . . .

But it is carrying coals to Newcastle to talk of such business arrangements as these to you.

The only thing I am strong about, is the folly of going on cutting blocks with our Secretarial razors any longer.

I am afraid I cannot give a very good account of myself.

The truth of the answer to Mallock's question " Is life worth living ? "—that depends on the liver—is being strongly enforced upon me in the hepatic sense of liver, and I must confess myself fit for very little. A week hence we shall migrate to Florence and try the effect of the more bracing air. The Pincio is the only part of Rome that is fit to live in, and unfortunately the Government does not offer to build me a house there.

However, I have got a great deal of enjoyment out of ancient Rome—papal Rome is too brutally pagan (and in the worst possible taste too) for me.

To his Daughter, Mrs. Roller

Jan. 11, 1885.

We have now had nearly three weeks in Rome. I am sick of churches, galleries, and museums, and meanly

make M—— go and see them and tell me about them.
As we are one flesh, it is just the same as if I had seen
them.

Since the time of Constantine there has been nothing
but tawdry rubbish in the shape of architecture [1]—the
hopeless bad taste of the Papists is a source of continual
gratification to me as a good Protestant (and something
more). As for the skies, they are as changeable as those
of England—the only advantage is the absence of frost
and snow—(raining cats and dogs this Sunday morning).

But down to the time of Constantine, Rome is endlessly
interesting, and if I were well I should like to spend
some months in exploring it. As it is, I do very little,
though I have contrived to pick up all I want to know
about Pagan Rome and the Catacombs, which last are my
especial weakness.

My master and physician is bothered a good deal with
eczema—otherwise very lively. All the chief collections
in Rome are provided with a pair of her spectacles, which
she leaves behind. Several new opticians' shops are set
up on the strength of the purchases in this line she is
necessitated to make.

I want to be back at work, but I am horribly afraid
I should be no good yet. We are thinking of going to
Florence at the end of this week to see what the drier
and colder air there will do.

With our dear love to you all—we are wae for a sight
of you—Ever your loving father,   T. H. HUXLEY.

HOTEL VICTORIA, VIA DEI DUE MACELLI,
Jan. 16, 1885.

MY DEAR FOSTER—It seems to me that I am giving
my friends a world of trouble. . . .

I have had a bad week of it, and the night before last

---

[1] For his appreciation of the great dome of the Pantheon, see
pp. 392, 442.

was under the impression that I was about to succumb
shortly to a complication of maladies, and moreover, that
a wooden box that my wife had just had made would
cost thousands of pounds in the way of payment for
extra luggage before we reached home. I do not know
which hypochondriacal possession was the most depressing.
I can laugh at it now, but I really was extraordinarily
weak and ill.

We had made up our minds to bolt from Rome to
Florence at once, when I suddenly got better, and to-day
am all right. So as we hear of snow at Florence we
shall stop where we are. It has been raining cats and
dogs here, and the Tiber rose 40 feet and inundated the
low grounds. But " cantabit elevatus "; it can't touch
us, and at any rate the streets are washed clean.

The climate is mild here. We have a capital room
and all the sunshine that is to be had, plus a good fire
when needful, and at worst one can always get a breezy
walk on the Pincio hard by.

However, about the leave. Am I to do anything or
nothing ? I am dying to get back to steady occupation
and English food, and the sort of regimen one can maintain
in one's own house. On the other hand, I stand in fear
of the bitter cold of February and early March, and still
more of the thousand and one worries of London outside
one's work. So I suppose it will be better if I keep away
till Easter, or at any rate to the end of March. But I
must hear something definite from the H.O. I have
written to Donnelly to the same effect. My poor Marian's
relapse did not do us any good, for all that I expected it.
However the last accounts are very favourable.

I wrote to Evans the other day about a re-arrange-
ment of the duties of the Secretary and Assistant Secretary.
I thought it was better to write to him than to you on
that subject, and I begged him to discuss the matter with
the officers. It is quite absurd that Stokes and you
should waste your time in press drudgery.

We are very prudent here, and the climate suits us both, especially my wife, who is so vigorous that I depute her to go and see the Palazzi, and tell me all about them when she comes back. Old Rome is endlessly interesting to me, and I can always potter about and find occupation. I think I shall turn antiquary—it's just the occupation for a decayed naturalist, though you need not tell the Treasurer I say so.

With our love to Mrs. Foster and yourself—Ever yours,   T. H. H.

HOTEL VICTORIA, ROME, VIA DEI DUE MACELLI,
Jan. 18, 1885.

MY DEAR DONNELLY—Official sentence of exile for two months more (up to May 12) arrived yesterday. So if my lords will be so kind as to concur I shall be able to disport myself with a clear conscience. I hope their lordships won't think that I am taking things too easy in not making a regular application, and I will do so if you think it better. But if it had rested with me I think I should have got back in February and taken my chance. That energetic woman that owns me, and Michael Foster, however, have taken the game out of my hands, and I have nothing to do but to submit.

On the whole I feel it is wise. I shall have more chance if I escape not only the cold but the bother of London for a couple of months more.

I was very bad a week ago, but I have taken to dosing myself with quinine, and either that or something else has given me a spurt for the last two days, so that I have been more myself than any time since I left, and begin to think that there is life in the old dog yet. If one could only have some fine weather! To-day there is the first real sunshine we have been favoured with for a week.

We are just back from a great function at St. Peter's,

It is the festa of St. Peter's chair, and the ex-dragoon Cardinal Howard has been fugleman in the devout adorations addressed to that venerable article of furniture, which, as you ought to know, but probably don't, is inclosed in a bronze double and perched up in a shrine of the worst possible taste in the Tribuna of St. Peter's. The display of man-millinery and lace was enough to fill the lightest-minded woman with envy, and a general concert—some of the music very good—prevented us from feeling dull, while the ci-devant guardsman—big, burly, and bullet-headed—made God and then eat him.[1] I must have a strong strain of Puritan blood in me somewhere, for I am possessed with a desire to arise and slay the whole brood of idolaters whenever I assist at one of these ceremonies. You will observe that I am decidedly better, and have a capacity for a good hatred still.

The last news about Gordon is delightful. The chances are he will rescue Wolseley yet.

With our love—Ever yours,   T. H. HUXLEY.

To his Eldest Son

ROME, *Jan.* 20, 1885.

I need hardly tell you that I find Rome wonderfully interesting, and the attraction increases the longer one stays. I am obliged to take care of myself and do but little in the way of sight-seeing, but by directing one's attention to particular objects one can learn a great deal without much trouble. I begin to understand Old Rome pretty well, and I am quite learned in the Catacombs, which suit me, as a kind of Christian fossils out of which one

---

[1] A reminiscence of Browning in "The Bishop Orders his Tomb ":—

And then how I shall lie through centuries,
And hear the blessed mutter of the mass,
And see God made and eaten all day long.

can reconstruct the body of the primitive Church. She was a simple maiden enough and vastly more attractive than the bedizened old harridan of the modern Papacy, so smothered under the old clothes of Paganism which she has been appropriating for the last fifteen centuries that Jesus of Nazareth would not know her if he met her.

I have been to several great papistical functions—among others to the festa of the Cathedra Petri in St. Peter's last Sunday, and I confess I am unable to understand how grown men can lend themselves to such elaborate tomfooleries—nothing but mere fetish worship —in forms of execrably bad taste, devised, one would think, by a college of ecclesiastical man-milliners for the delectation of school-girls. It is curious to notice that intellectual and æsthetic degradation go hand in hand. You have only to go from the Pantheon to St. Peter's to understand the great abyss which lies between the Roman of paganism and the Roman of the papacy. I have seen nothing grander than Agrippa's work—the popes have stripped it to adorn their own petrified lies, but in its nakedness it has a dignity with which there is nothing to compare in the ill-proportioned, worse decorated tawdry stone mountain on the Vatican.

The best thing, from an æsthetic point of view, that could be done with Rome would be to destroy everything except St. Paolo fuori le Mura, of later date than the fourth century.

But you will have had enough of my scrawl, and your mother wants to add something. She is in great force, and is gone prospecting to some Palazzo or other to tell me if it is worth seeing.—Ever your loving father,

T. H. HUXLEY.

HOTEL VICTORIA, ROME, VIA DEI DUE MACELLI,
Jan. 25, 1885.

MY DEAR DONNELLY—Best thanks for the telegram which arrived the day before yesterday and set my mind at ease.

I have been screwing up the old machine which I inhabit, first with quinine and now with a form of strychnia (which Clark told me to take) for the last week, and I have improved a good deal whether *post hoc* or *propter hoc* in the present uncertainty of medical science I decline to give any opinion.

The weather is very cold for Rome—ice an eighth of an inch thick in the Ludovisi Garden the other morning, and every night it freezes, but mostly fine sunshine in the day. (This is a remarkable sentence in point of grammar, but never mind.) The day before yesterday we came out on the Campagna, and it then was as fresh and bracing a breeze as you could get in Northumberland.

We are very comfortable and quiet here, and I hold on—till it gets warmer. I am told that Florence is detestable at present. As for London, our accounts make us shiver and cough.

News about the dynamiting gentry just arrived. A little more mischief and there will be an Irish massacre in some of our great towns. If an Irish Parnellite member were to be shot for every explosion I believe the thing would soon stop. It would be quite just, as they are practically accessories.

I think——would do it if he were Prime Minister. Nothing like a thorough Radical for arbitrary acts of power !

I must be getting better, as my disgust at science has ceased, and I have begun to potter about Roman geology and prehistoric work. You may be glad to learn that there is no evidence that the prehistoric Romans had Roman noses. But as I cannot find any particular prevalence of [them] among the modern—or ancient except for Cæsar—Romani, the fact is not so interesting as it might appear, and I would not advise you to tell —— of it.

Behold a Goak—feeble, but promising of better things.

My wife unites with me with love to Mrs. Donnelly and yourself.—Ever yours,   T. H. HUXLEY.

The following letter refers to the fourth edition of the *Lessons in Elementary Physiology*, in the preparation of which Dr. Foster had been helping during the summer :—

HOTEL VICTORIA, ROME, VIA DEI DUE MACELLI
Feb. 1, 1885.

MY DEAR FOSTER—Anything more disgraceful than the way in which I have left your letter of more than a fortnight ago unanswered, I don't know. I thought the wife had written about the leave (and she thought I had, as she has told you), but I knew. I had not answered the questions about the title, still less considered the awful incubus (× 10,000 dinners by hepatic deep objection) of the preface.

There is such a thing as justice in this world—not much of it, but still some—and it is partly on that ground and partly because I want you, in view of future eventualities, to have a copyright in the book, that I proposed we should join our names.

Of course, if you would really rather not, for any good reason you may have, I have nothing further to say. But I don't think that the sentimental reason is a good one, and unless you have a better, I wish you would let the original proposal stand.

However, having stated the case afresh I leave it for you to say yes or no, and shall abide by your decision without further discussion.

As to the Preface. If I am to write it, please send
me the old Preface. I think the book was published in
1864, or was it 1866 ?[1] and it ought to be come of age
or nearly so.

You might send me the histological chapter, not that
I am going to alter anything, but I should like to see
how it looks. I will knock the Preface off at once, as
soon as I hear from you.

The fact is, I have been much better in the course of
the last few days. The weather has been very sunshiny
but cool and bracing, and I have taken to quinine.
Tried Clark's strychnine, but it did not answer so well.

I am in hopes that I have taken a turn for the better,
and that there may yet be the making of something
better than a growling hypochondriacal old invalid about
me. But I am most sincerely glad that I am not obliged
to be back 10 days hence—there is not much capital
accumulated yet.

I find that the Italians have been doing an immense
deal in prehistoric archæology of late years, and far more
valuable work than I imagined. But it is very difficult
to get at, and as Loescher's head man told me the other
day when I asked for an Italian book published in Rome,
"Well, you see it is so difficult to get Roman books in
Rome."

I am ashamed to be here two months without paying
my respects to the Lincei, and I am going to-day. The
unaccountable creatures meet at 1 o'clock—lunch time !

Best love from my wife and self to Mrs. Foster and
yourself.—Ever yours,   T. H. HUXLEY.

ROME, *Feb.* 14, 1885.

MY DEAR FOSTER.—*Voilà* the preface—a work of
great labour ! and which you may polish and alter as you

---

[1] In 1866.

like, *all but the last paragraph.* You see I have caved in. I like your asking to have your own way "for once." My wife takes the same line, does whatever she pleases, and then declares I leave her no initiative.

If I talk of public affairs, I shall simply fall a-blaspheming. I see the *Times* holds out about Gordon, and does not believe he is killed. Poor fellow! I wish I could believe that his own conviction (as he told me) is true, and that death only means a larger government for him to administer. Anyhow, it is better to wind up that way than to go growling out one's existence as a ventose hypochondriac, dependent upon the condition of a few square inches of mucous membrane for one's heaven or hell.

As to private affairs, I think I am getting solidly, but very slowly, better. In fact, I can't say there is much the matter with me, except that I am weaker than I ought to be, and that a sort of weary indolence hangs about me like a fog. M—— is wonderfully better, and her husband has taken a house for them at Norwood. If I could be rejoiced at anything, I should be at that; but it seems to me as if since that awful journey when I first left England, "the springs was broke," as that vagabond tout said at Naples.

It has turned very cold here, and we are uncertain when to leave for Florence, but probably next week. The Carnival is the most entirely childish bosh I have ever met with among grown people. Want to finish this now for post, but will write again speedily. Moseley's proposition is entirely to my mind, and I have often talked of it. The R.S. rooms ought to be house-of-call and quasi-club for all F.R.S. in London.

Wife is bonny, barring a cold. It is as much as I can do to prevent her sporting a mask and domino!

With best love—Ever yours,

T. H. H.

HOTEL VICTORIA, ROME,
VIA DEI DUE MACELLI, *Feb.* 16, 1885.

MY DEAR DONNELLY—I have had it on my mind to write to you for the last week—ever since the hideous news about Gordon reached us. But partly from a faint hope that his wonderful fortune might yet have stood him in good stead, and partly because there is no great satisfaction in howling with rage, I have abstained.

Poor fellow ! I wonder if he has entered upon the "larger sphere of action" which he told me was reserved for him in case of such a trifling accident as death. Of all the people whom I have met with in my life, he and Darwin are the two in whom I have found something bigger than ordinary humanity—an unequalled simplicity and directness of purpose—a sublime unselfishness.

Horrible as it is to us, I imagine that the manner of his death was not unwelcome to himself. Better wear out than rust out, and better break than wear out. The pity is that he could not know the feeling of his countrymen about him.

I shall be curious to see what defence the super-ingenious Premier has to offer for himself in Parliament. I suppose, as usual, the question will drift into a brutal party fight, when the furious imbecility of the Tories will lead them to spoil their case. That is where we are ; on the one side, timid imbecility "waiting for instructions from the constituencies" ; furious imbecility on the other, looking out for party advantage. Oh ! for a few months of William Pitt.

I see you think there may be some hope that Gordon has escaped yet. I am afraid the last telegram from Wolseley was decisive. We have been watching the news with the greatest anxiety, and it has seemed only to get blacker and blacker.

．   ．   ．   ．   ．

[Touching a determined effort to alter the management of certain Technical Education business.]

I trust he may succeed, and that the unfitness of these people to be trusted with anything may be demonstrated. I regret I am not able to help in the good work. Get the thing out of their hands as fast as possible. The prospect of being revenged for all the beastly dinners I sat out and all the weary discussions I attended to no purpose, really puts a little life into me. Apropos of that, I am better in various ways, but curiously weak and washed out; and I am afraid that not even the prospect of a fight would screw me up for long. I don't understand it, unless I have some organic disease of which nobody can find any trace (and in which I do not believe myself), or unless the terrible trouble we have had has accelerated the advent of old age. I rather suspect that the last speculation is nearest the truth. You will be glad to hear that my poor girl is wonderfully better, and, indeed, to all appearance quite well. They are living quietly at Norwood.

I shall be back certainly by the 12th April, probably before. We have found very good quarters here, and have waited for the weather to get warmer before moving; but at last we have made up our minds to begin nomadising again next Friday. We go to Florence, taking Siena, and probably Pisa, on our way, and reaching Florence some time next week. Address—Hotel Milano, Via Cerretani.

For the last week the Carnival has been going on. It strikes me as the most elaborate and dreariest tomfoolery I have ever seen, but I doubt if I am in the humour to judge it fairly. It is only just to say that it entertains my vigorous wife immensely. I have been expecting to see her in mask and domino, but happily this is the last day, and there is no sign of any yet. I have never seen any one so much benefited by rest and change as she is, and that is a good thing for both of us.

After Florence we shall probably make our way to Venice, and come home by the Lago di Garda and

Germany. But I will let you know when our plans are settled.

With best love from we two to you two—Ever yours,

T. H. HUXLEY.

To his Youngest Daughter

SIENA, *Feb.* 23, 1885.

DEAREST ETHEL—The cutting you sent me contains one of the numerous "goaks" of a Yankee performing donkey who is allowed to disport himself in one of the New York papers. I confess it is difficult to see the point of the joke, but there is one if you look close. I don't think you need trouble to enlighten the simple inquirer. He probably only wanted the indignant autograph which he won't get.

The Parker Museum must take care of itself. The public ought to support it, not the men of science.

As a grandfather, I am ashamed of my friends who are of the same standing; but I think they would take it as a liberty if, in accordance with your wish, I were to write to expostulate.

After your mother had exhausted the joys of the Carnival, she permitted me to leave Rome for this place, where we arrived last Friday evening. My impression is that if we had stayed in Rome much longer we should never have left. There is something idle and afternoony about the air which whittles away one's resolution.

The change here is wonderfully to the good. We are perched more than a thousand feet above the sea, looking over the Tuscan hills for twenty or thirty miles every way. It is warm enough to sit with the window wide open and yet the air is purer and more bracing than in any place we have visited. Moreover, the hotel (Grande Albergo) is very comfortable.

Then there is one of the most wonderful cathedrals to

be seen in all North Italy—free from all the gaudy finery and atrocious bad taste which have afflicted me all over South Italy. The town is the quaintest place imaginable —built of narrow streets on several steep hills to start with, and then apparently stirred up with a poker to prevent monotony of effect.

Moreover, there is Catherine of Siena, of whom I am reading a delightful Catholic life by an Italian father of the Oratory. She died 500 years ago, but she was one of twenty-five children, and I think some of them must have settled in Kent and allied themselves with the Heathorns. Otherwise, I don't see why her method of writing to the Pope should have been so much like the way my daughters (especially the youngest) write to their holy father.

I wish she had not had the stigmata—I am afraid there must have been a *leetle* humbug about the business —otherwise she was a very remarkable person, and you need not be ashamed of the relationship.

I suppose we shall get to Florence some time this week ; the address was sent to you before we left Rome —Hotel Milano, Via Cerretani. But I am loth to leave this lovely air in which, I do believe, I am going to pick up at last. The misfortune is that we did not intend to stay here more than three days, and so had letters sent to Florence. Everybody told us it would be very cold, and, as usual, everybody told taradiddles.

M—— unites in fondest love to you all.—Ever your loving father,   T. H. HUXLEY.

To HIS SON

SIENA, *Feb.* 25, 1885.

. . . If you had taken to physical science it would have been delightful to me for us to have worked together, and I am half inclined to take to history that I may earn that pleasure. I could give you some capital wrinkles about the physical geography and prehistoric history (excuse bull) of Italy for a Roman History primer! Joking apart, I believe that history might be, and ought to be, taught in a new fashion so as to make the meaning of it as a process of evolution—intelligible to the young. The Italians have been doing wonders in the last twenty years in prehistoric archæology, and I have been greatly interested in acquainting myself with the general results of their work.

We moved here last Friday, and only regret that the reports of the weather prevented us from coming sooner. More than 1000 ft. above the sea, in the midst. of a beautiful hill country, and with the clearest and purest air we have met with in Italy, Siena is perfectly charming. The window is wide open and I look out upon a vast panorama, something like that of the Surrey hills, only on a larger scale—" Raw Siena," "Burnt Siena," in the foreground, where the colour of the soil is not hidden by the sage green olive foliage, purple mountains in the distance.

The old town itself is a marvel of picturesque crookedness, and the cathedral a marvel. M. and I have been devoting ourselves this morning to St. Catarina and Sodoma's pictures.

I am reading a very interesting life of her by Capecelatro, and, if my liver continues out of order, may yet turn Dominican.

However, the place seems to be doing me good, and I may yet, like another person, decline to be a monk.

To his Daughter, Mrs. Roller

March 8.

The great merit of Rome is that you have never seen the end of it. M. and I have not worked very hard at our galleries and churches, but I have got so far as a commencing dislike for the fine arts generally. Perhaps after a week or two I shall take to science out of sheer weariness.

HOTEL DE MILANO, FLORENCE,
March 12, 1885.

MY DEAR FOSTER—My wife and I send you our hearty good wishes (antedated by four days). I am not sure we ought not to offer our best thanks to your mother for providing us with as staunch a friend as people ever were blessed with. It is possible that she did not consider that point nine and forty years ago ; but we are just as grateful as if she had gone through it all on our own account.

We start on our way homeward to-morrow or next day, by Bologna to Venice, and then to England by the way we came—taking it easy. The Brenner is a long way round and I hear very cold. I think we may stay a few days at Lugano, which I liked very much when there before. Florence is very charming, but there is not much to be said for the climate. My wife has been bothered with sore throat, to which she is especially liable, ever since we have been here. Old residents console her with the remark that Florentine sore throat is a regular thing in the spring. The alternations of heat and cold are detestable. So we stand thus—*Naples*, bad for both—*Rome*, good for her, bad for me—*Florence*, bad for her, baddish for me. Venice has to be tried, but stinks and mosquitoes are sure to render it impossible as soon as the weather is warm. Siena is the only place that suited both of us, and I don't think that would exactly answer to live in. Nothing like foreign travel for making one content with home.

I shall have to find a country lot suited to my fortunes when I am paid off. Couldn't you let us have your gardener's cottage ? My wife understands poultry and

I shall probably have sufficient strength to open the gate and touch my hat to the Dons as they drive up. I am afraid E. is not steady enough for waiting-maid or I would offer her services.

. . . I am rejoiced to hear that the lessons [1] and the questions are launched. They loom large to me as gigantic undertakings, in which a dim and speculative memory suggests I once took part, but probably it is a solar myth, and I am too sluggish to feel much compunction for the extra trouble you have had.

Perhaps I shall revive when my foot is on my native heath in the shady groves of the Evangelist.[2]

My wife is out photograph hunting—nothing diminishes her activity—otherwise she would join in love and good wishes to Mrs. Foster and yourself.—Ever yours,

T. H. HUXLEY.

The two worst and most depressing periods of this vain pilgrimage in pursuit of health were the stay at Rome and at Florence. At the latter town he was inexpressibly ill and weak; but his daily life was brightened by the sympathy and active kindness of Sir Spencer Walpole, who would take him out for short walks, talking as little as possible, and shield him from the well-meant but tactless attentions of visitors who would try to "rouse him and do him good" by long talks on scientific questions.

His physical condition, indeed, was little improved.

As for my unsatisfactory carcase (he writes on March 6, to Sir J. Donnelly), there seems nothing the matter with it now except that the brute objects to work. I eat well, drink well, sleep well, and have no earthly ache,

---

[1] The new edition of the *Elementary Physiology*.
[2] St. John's Wood.

pain or discomfort. I can walk for a couple of hours or more without fatigue. But half an hour's talking wearies me inexpressibly, and "saying a few words," would finish me for the day. For all that, I do not mean to confess myself finally beaten till I have had another try.

That is to say, he was still bent upon delivering his regular course of lectures at South Kensington as soon as he returned, in spite of the remonstrances of his wife and his friends.

In the same letter he contrasts Florence with Siena and its "fresh, elastic air," its "lovely country that reminds one of a magnified version of the Surrey weald." The Florentine climate was trying.[1] "And then there is the awful burden of those miles of 'treasures of art.'" He had been to the Uffizii; "and there is the Pitti staring me in the face like drear fate. Why can't I have the moral courage to come back and say I haven't seen it? I should be the most distinguished of men."

There is another reference to Gordon :—

What an awful muddle you are all in in the bright little, tight little island. I hate the sight of the English

---

[1] A week later he writes to Sir J. Evans—"I begin to look forward with great satisfaction to the equability of English weather —to that dear little island where doors and windows shut close— where fires warm without suffocating—where the chief business of the population in the streets is something else than expectoration —and where I shall never see fowl with salad again.

"You perceive I am getting better by this prolonged growl. . . . But half an hour's talking knocks me up, and I am such an effete creature that I think of writing myself p. R. S. with a small p."

papers. The only good thing that has met my eye lately
is a proposal to raise a memorial to Gordon. I want to
join in whatever is done, and unless it will be time
enough when I return, I shall be glad if you will put
me down for £5 to whatever is the right scheme.

The following to his daughter, Mrs. Roller, describes the stay in Florence.

HOTEL DE MILANO, FLORENCE,
March 7, 1885.

We have been here more than a week and have
discovered two things, first that the wonderful "art
treasures," of which all the world has heard, are a sore
burden to the conscience if you don't go to see them, and
an awful trial to the back and legs if you do ; and thirdly,
that the climate is productive of a peculiar kind of relaxed
throat. M.'s throat discovered it, but on inquiry, it
proved to be a law of nature, at least, so the oldest in-
habitants say. We called on them to-day.

But it is a lovely place for all that, far better than
Rome as a place to live in, and full of interesting things.
We had a morning at the Uffizii the other day, and
came back with minds enlarged and backs broken. To-
morrow we contemplate attacking the Pitti, and doubt
not the result will be similar. By the end of the week
our minds will probably [be] so large, and the small of
the back so small that we should probably break if we
stayed any longer, so think it prudent to be off to Venice.
Which Friday is the day we go, reaching Venice Saturday
or Sunday. Pension Suisse, Canal Grande, as before.
And mind we have letters waiting for us there, or your
affectionate Pater will emulate the historical " cocky."

I got much better at Siena, probably the result of the
medicinal nature of the city, the name of which, as a
well-instructed girl like you knows, is derived from the

senna, which grows wild there, and gives the soil its peculiar pigmentary character.

But unfortunately I forgot to bring any with me, and the effect went off during the first few days of our residence here, when I was, as the Italians say, "molto basso nel bocca." However I am picking up again now, and if people wouldn't call upon us, I feel there might be a chance for me.

I except from that remark altogether the dear Walpoles who are here and as nice as ever. Mrs. Walpole's mother and sister live here, and the W.'s are on a visit to them but leave on Wednesday. They go to Venice, but only for two or three days.

We shall probably stay about a fortnight in Venice, and then make our way back by easy stages to London. We are wae to see you all again.

Doctor M—— [Mrs. Huxley] has just been called in to a case of sore throat in the person of a young lady here, and is quite happy. The young lady probably will not be, when she finds herself converted into a sort of inverted mustard-pot, with the mustard outside! She is one of a very nice family of girls, who (by contrast) remind us of own.—Ever your loving (to all) father, PATER.

Mrs. M—— has just insisted on seeing this letter.

To His Youngest Daughter

HOTEL BEAU SÉJOUR, SAN REMO,
March 30, 1885.

DEAREST BABS—We could not stand "beautiful Venice the pride of the sea" any longer. It blew and rained and colded for eight-and-forty hours consecutively. Everybody said it was a most exceptional season, but that did not make us any warmer or prevent your mother from catching an awful cold. So as soon as she got better we packed up and betook ourselves here by way of Milan and Genoa. At Milan it was so like London on a wet day, that except for the want of smoke we might have been in our dear native land. At Genoa we arrived late one afternoon and were off early in the morning—but by dint of taking a tram after dinner (not a dram) and going there and back again we are able to say we have seen that city of palaces. The basements we saw through the tram windows by mixed light of gas and moon may in fact all have belonged to palaces. We are not in a position to say they did not.

The quick train from Genoa here is believed to go fully twenty-five miles an hour, but starts at 7 A.M., but the early morning air being bad for the health, we took the slow train at 9.30, and got here some time in the afternoon. But mind you it is a full eighty miles, and when we were at full speed between the stations—very few donkeys could have gone faster. But the coast scenery is very pretty, and we didn't mind.

Here we are very well off and as nearly warm as I expect to be before reaching England. You can sit out in the sun with satisfaction, though there is a little knife-edge of wind just to remind us of Florence. Everybody, however, tells us it is quite an exceptional season, and that it ought to be the most balmy air imaginable. Besides there are no end of date-palms and cactuses and aloes and odorous flowers in the garden—and the loveliest purple sea you can imagine.

Well, we shall stop some days and give San Remo a chance—at least a week, unless the weather turns bad.

As to your postcards which have been sent on from Venice and are really shabby, I am not going to any dinners whatsoever, either Middle Temple or Academy. Just write to both that "Mr. H. regrets he is unable to accept the invitation with which——have honoured him."[1]

---

[1] "It's like putting the shutters up," he said sadly to his wife, when he felt unable to attend the Royal Academy dinner as he had done for many years.

I have really nothing the matter with me now—but my stock of strength is not great, and I can't afford to spend any on dinners.

The blessedest thing now will be to have done with the nomadic life of the last five months—and see your ugly faces (so like their dear father) again. I believe it will be the best possible tonic for me.

M——has not got rid of her cold yet, but a few warm days here will, I hope, set her up.

I met Lady Whitworth on the esplanade to-day—she is here with Sir Joseph, and this afternoon we went to call on her. The poor old man is very feeble and greatly altered since I saw him last.

Write here on receiving this. We shall take easy stages home, but I don't know that I shall be able to give you any address.

M——sends heaps of love to all (including Charles [1]) —Ever your loving father,   T. H. HUXLEY.

Tell the "Micropholis" man that it is a fossil lizard with an armour of small scales.

---

[1] The cat.

CHAPTER XVII

1885

ON April 8, he landed at Folkestone, and stayed
there a day or two before going to London. Writing
to Sir J. Donnelly, he remarks with great satisfaction
at getting home :—

> We got here this afternoon after a rather shady
> passage from Boulogne, with a strong north wind in our
> teeth all the way, and rain galore. For all that, it is
> the pleasantest journey I have made for a long time—so
> pleasant to see one's own dear native mud again. There
> is no foreign mud to come near it.

And on the same day he sums up to Sir M. Foster
the amount of good he has gained from his expedition, and the amount of good any patient is likely to
get from travel :—

> As for myself I have nothing very satisfactory to say.
> By the oddest chance we met Andrew Clark in the boat,
> and he says I am a very bad colour—which I take it is
> the outward and visible sign of the inward and carnal
> state. I may sum that up by saying that there is
> nothing the matter but weakness and indisposition to do

anything, together with a perfect genius for making mountains out of molehills.

After two or three fine days at Venice, we have had nothing but wet or cold—or hot and cold at the same time, as in that prodigious imposture the Riviera. Of course it was the same story everywhere, "perfectly unexampled season."

Moral.—If you are perfectly well and strong, brave Italy—but in search of health stop at home.

It has been raining cats and dogs, and Folkestone is what some people would call dreary. I could go and roll in the mud with satisfaction that it is English mud.

It will be jolly to see you again. Wife unites in love.—Ever yours,   T. H. HUXLEY.

To return home was not only a great pleasure; it gave him a fillip for the time, and he writes to Sir M. Foster, April 12 :—

It is very jolly to be home, and I feel better already. Clark has just been here overhauling me, and feels very confident that he shall screw me up.

I have renounced dining out and smoking (!!!) by way of preliminaries. God only knows whether I shall be permitted more than the smell of a mutton chop for dinner. But I have great faith in Andrew, who set me straight before when other "physicians' aid was vain."

But his energy was fitful; lassitude and depression again invaded him. He was warned by Sir Andrew Clark to lay aside all the burden of his work. Accordingly, early in May, just after his sixtieth birthday, he sent in his formal resignation of the Professorship of Biology, and the Inspectorship of Salmon Fisheries; while a few days later he laid his resignation of the

Presidency before the Council of the Royal Society. By the latter he was begged to defer his final decision, but his health gave no promise of sufficient amendment before the decisive Council meeting in October. He writes on May 27 :—

I am convinced that what with my perennial weariness and my deafness I ought to go, whatever my kind friends may say.

A curious effect of his illness was that for the first time in his life he began to shrink involuntarily from assuming responsibilities and from appearing on public occasions ; thus he writes on June 16 :—

I am sorry to say that the perkiness of last week [1] was only a spurt, and I have been in a disgusting state of blue devils lately. Can't make out what it is, for I really have nothing the matter, except a strong tendency to put the most evil construction upon everything.

I am fairly dreading to-morrow [*i.e.* receiving the D.C.L. degree at Oxford [2]] but why I don't know— probably an attack of modesty come on late in life and consequently severe.

Very likely it will do me good and make me "fit" for Thursday [*i.e.* Council and ordinary meetings of Royal Society].

And a month later :

I have been idling in the country for two or three days—but like the woman with the issue, "I am not better but rather worse"—blue devils and funk—funk

---

[1] *I.e.* at the unveiling of the Darwin statue at South Kensington. See p. 422.
[2] See p. 419.

and blue devils. Liver, I expect. [An ailment of which he says to Prof. Marsh, "I rather wish I had some respectable disease—it would be livelier."]

And again :—

Everybody tells me I look so much better, that I am really ashamed to go growling about, and confess that I am continually in a blue funk and hate the thought of any work—especially of scientific or anything requiring prolonged attention.

At the end of July he writes to Sir W. Flower—

4 MARLBOROUGH PLACE,
July 27, 1885.

MY DEAR FLOWER—I am particularly glad to hear that things went right on Saturday, as my conscience rather pricked me for my desertion of the meeting.[1] But it was the only chance we had of seeing our young married couple before the vacation—and you will rapidly arrive at a comprehension of the cogency of *that* argument now.

I will think well of your kind words about the Presidency. If I could only get rid of my eternal hypochondria the work of the R.S. would seem little enough. At present, I am afraid of everything that involves responsibility to a degree that is simply ridiculous. I only wish I could shirk the inquiries I am going off to hold in Devonshire !

P.R.S. in a continual blue funk is not likely to be either dignified or useful ; and unless I am in a better frame of mind in October I am afraid I shall have to go. —Ever yours very faithfully,   T. H. HUXLEY.

---

[1] British Museum Trustees, July 25.

A few weeks at Filey in August did him some
good at first; and he writes cheerfully of his lodgings
in "a place with the worst-fitting doors and windows,
and the hardest chairs, sofas, and beds known to my
experience."

He continues :—

I am decidedly picking up. The air here is wonderful, and as we can set good cookery against hard lying (I
don't mean in the Munchausen line) the consequent
appetite becomes a mild source of gratification. Also, I
have not met with more than two people who knew me,
and that in my present state is a negative gratification
of the highest order.

Later on he tried Bournemouth; being no better,
he thought of an entirely new remedy.

The only thing I am inclined to do is to write a book
on Miracles. I think it might do good and unload my
biliary system.

In this state of indecision, so unnatural to him, he
writes to Sir M. Foster :—

I am anything but clear as to the course I had best
take myself. While undoubtedly much better in general
health, I am in a curious state of discouragement, and I
should like nothing better than to remain buried here
(Bournemouth) or anywhere else, out of the way of
trouble and responsibility. It distresses me to think
that I shall have to say something definite about the
Presidency at the meeting of Council in October.

Finally on October 20, he writes :—

I think the lowest point of my curve of ups and
downs is gradually rising—but I have by no means

reached the point when I can cheerfully face anything.
I got over the Board of Visitors (two hours and a half)
better than I expected, but my deafness was a horrid
nuisance.

I believe the strings of the old fiddle will tighten up
a good deal, if I abstain from attempting to play upon
the instrument at present—but that a few jigs now will
probably ruin that chance.

But I will say my final word at our meeting next
week. I would rather step down from the chair than
dribble out of it. Even the devil is in the habit of de-
parting with a "melodious twang," and I like the
precedent.

So at the Anniversary meeting on November 30, he
definitely announced in his last Presidential address
his resignation of that "honourable office" which he
could no longer retain "with due regard to the
interests of the Society, and perhaps, I may add, of
self-preservation."

I am happy to say (he continued) that I have good
reason to believe that, with prolonged rest—by which I
do not mean idleness, but release from distraction and
complete freedom from those lethal agencies which are
commonly known as the pleasures of society—I may yet
regain so much strength as is compatible with advancing
years. But in order to do so, I must, for a long time
yet, be content to lead a more or less anchoritic life.
Now it is not fitting that your President should be a
hermit, and it becomes me, who have received so much
kindness and consideration from the Society, to be particularly careful that no sense of personal gratification
should delude me into holding the office of its representative one moment after reason and conscience have pointed
out my incapacity to discharge the serious duties which

devolve upon the President, with some approach to efficiency.

I beg leave, therefore, with much gratitude for the crowning honour of my life which you have conferred upon me, to be permitted to vacate the chair of the Society as soon as the business of this meeting is at an end.

The settlement of the terms of the pension upon which, after thirty-one years of service under Government, he retired from his Professorship at South Kensington and the Inspectorship of Fisheries, took a considerable time. The chiefs of his own depart ment, that of Education, wished him to retire upon full pay, £1500 (see p. 289). The Treasury were more economical. It was the middle of June before the pension they proposed of £1200 was promised ; the end of July before he knew what conditions were attached to it.

On June 20, he writes to Mr. Mundella, Vice-President of the Council :—

MY DEAR MUNDELLA—Accept my warmest thanks for your good wishes, and for all the trouble you have taken on my behalf. I am quite .ashamed to have been the occasion of so much negotiation.

Until I see the Treasury letter, I am unable to judge what the £1200 may really mean,[1] but whatever the result, I shall never forget the kindness with which my chiefs have fought my battle.—I am, yours very faithfully,

T. H. HUXLEY.

---

[1] *I.e.* whether he was to draw his salary of £200 as Dean or not.

On July 16, he writes to Sir M. Foster :—

The blessed Treasury can't make up their minds whether I am to be asked to stay on as Dean or not, and till they do, I can't shake off any of my fetters.

Early in the year he had written to Sir John Donnelly of the necessity of resigning :—

Nevertheless (he added), it will be a sad day for me when I find myself no longer entitled to take part in the work of the schools in which you and I have so long been interested.

But that "sad day" was not to come yet. His connection with the Royal College of Science was not entirely severed. He was asked to continue, as Honorary Dean, a general supervision of the work he had done so much to organise, and he kept the title of Professor of Biology, his successors in the practical work of the chair being designated Assistant Professors.

"I retain," he writes, "general superintendence as part of the great unpaid."

It is a comfort (he writes to his son) to have got the thing settled. My great desire at present is to be idle, and I am now idle with a good conscience.

Later in the year, however, a change of Ministry having taken place, he was offered a Civil List Pension of £300 a year by Lord Iddesleigh. He replied accepting it :—

4 MARLBOROUGH PLACE,
Nov. 24, 1885.

MY DEAR LORD IDDESLEIGH—Your letters of the 20th November reached me only last night, and I hasten to thank you for both of them. I am particularly obliged for your kind reception of what I ventured to say about the deserts of my old friend Sir Joseph Hooker.

With respect to your Lordship's offer to submit my name to Her Majesty for a Civil List Pension, I can but accept a proposal which is in itself an honour, and which is rendered extremely gratifying to me by the great kindness of the expressions in which you have been pleased to embody it.

I am happy to say that I am getting steadily better at last, and under the regime of "peace with honour" that now seems to have fallen to my lot, I may fairly hope yet to do a good stroke of work or two.—I remain, my dear Lord Iddesleigh, faithfully yours,

T. H. HUXLEY.

4 MARLBOROUGH PLACE,
Nov. 24, 1885.

MY DEAR DONNELLY—I believe you have been at work again !

Lord Iddesleigh has written to me to ask if I will be recommended for a Civil List Pension of £300 a year, a very pretty letter, not at all like the Treasury masterpiece you admired so much.

Didn't see why I should not accept, and have accepted accordingly. When the announcement comes out the Liberals will say the Tory Govt. have paid me for attacking the G.O.M. ! to a dead certainty.—Ever yours,

T. H. HUXLEY.

Five days later he replies to the congratulations

of Mr. Eckersley (whose son had married Huxley's third daughter) :—

> . . . Lord Iddesleigh's letter offering to submit my name for an honorary pension was a complete surprise.
>
> My chiefs in the late Government wished to retire me on full pay, but the Treasury did not see their way to it, and cut off £300 a year. Naturally I am not sorry to have the loss made good, but the way the thing was done is perhaps the pleasantest part of it.

There was a certain grim appropriateness in his "official death" following hard upon his sixtieth birthday, for sixty was the age at which he had long declared that men of science ought to be strangled, lest age should harden them against the reception of new truths, and make them into clogs upon progress, the worse, in proportion to the influence they had deservedly won. This is the allusion in a birthday letter from Sir M. Foster :—

> REVEREND SIR—So the "day of strangulation" has arrived at last, and with it the humble petition of your friends that you may be induced to defer the "happy dispatch" for, say at least ten years, when the subject may again come up for consideration. For your petitioners are respectfully inclined to think that if your sixtyship may be induced so far to become an apostle as to give up the fishery business, and be led to leave the Black Board at S.K. to others, the t'other side sixty years, may after all be the best years of your life. In any case they would desire to bring under your notice the fact that *they feel they want you as much as ever they did*.—Ever thine,   M. F.

Reference has been made to the fact that the honorary degree of D.C.L. was conferred this May upon Huxley by the University of Oxford. The Universities of the sister kingdoms had been the first thus to recognise his work; and after Aberdeen and Dublin, Cambridge, where natural science had earlier established a firm foothold, showed the way to Oxford. Indeed, it was not until his regular scientific career was at an end, that the University of Oxford opened its portals to him. So, as he wrote to Professor Bartholomew Price on May 20, in answer to the invitation, "It will be a sort of apotheosis coincident with my official death, which is imminent. In fact, I am dead already, only the Treasury Charon has not yet settled the conditions upon which I am to be ferried over to the other side."

Before leaving the subject of his connection with the Royal Society, it may be worth while to give a last example of the straightforward way in which he dealt with a delicate point whether to vote or not to vote for his friend Sir Andrew Clark, who had been proposed for election to the Society. It occurred just after his return from abroad; he explains his action to Sir Joseph Hooker, who had urged caution on hearing a partial account of the proceedings.

SOUTH KENSINGTON,
April 25, 1885.

MY DEAR HOOKER—I don't see very well how I could have been more cautious than I have been. I knew nothing of Clark's candidature until I saw his name in the list; and if he or his proposer had consulted me, I should have advised delay, because I knew very well there would be a great push made for —— this year.

Being there, however, it seemed to me only just to say that which is certainly true, namely, that Clark has just the same claim as half a dozen doctors who have been admitted without question, e.g. Gull, Jenner, Risdon Bennett, on the sole ground of standing in the profession. And I think that so long as that claim is admitted, it will be unjust not to admit Clark.

So I said what you heard; but I was so careful not to press unduly upon the Council, that I warned them of the possible prejudice arising from my own personal obligations to Clark's skill, and I went so far as not to put his name in the *first* list myself, a step which I now regret.

If this is not caution enough, I should like to know what is? As Clive said when he came back from India, " By God, sir, I am astonished at my own moderation!"

If it is not right to make a man a fellow because he holds a first-class place as a practitioner of medicine as the R.S. has done since I have known it, let us abolish the practice. But then let us also in justice refuse to recognise the half-and-half claims, those of the people who are third-rate as practitioners, and hang on to the skirts of science without doing anything in it.

Several of your and my younger scientific friends are bent on bringing in their chum ——, and Clark's candidature is very inconvenient to them. Hence I suspect some of the " outspoken aversion " and criticism of Clark's claims you have heard.

I am quite willing to sacrifice my friend for a principle, but not for somebody else's friend, and I mean to vote for Clark; though I am not going to try to force my notion down any one else's throat.—Ever yours faithfully,

T. H. HUXLEY.

On the same subject he writes to Sir M. Foster :—

Obedience be hanged. It would not lie in my mouth, as the lawyers say, to object to anybody's getting his own way if he can.

If Clark had not been a personal friend of mine I should not have hesitated a moment about deciding in his favour. Under the circumstances it was quite clear what I should do if I were forced to decide, and I thought it would have been kindly and courteous to the President if he had been let off the necessity of making a decision which was obviously disagreeable to him.

If, on the other hand, it was wished to fix the responsibility of what happened on him, I am glad that he had the opportunity of accepting it. I never was more clear as to what was the right thing to do.

So also at other times ; he writes in September to Sir M. Foster, the Secretary, with reference to evening gatherings at which smoking should be permitted.

BOURNEMOUTH, *Sept.* 17, 1885.

I am not at all sure that I can give my blessing to the "Tabagie." When I heard of it I had great doubts as to its being a wise move. It is not the question of "smoke" so much, as the principle of having meetings in the Society's rooms, which are not practically (whatever they may be theoretically), open to all the fellows, and which will certainly be regarded as the quasi-private parties of one of the officers. You will have all sorts of jealousies roused, and talk of a clique, etc.

When I was Secretary the one thing I was most careful to avoid was the appearance of desiring to exert any special influence. But there was a jealousy of the x Club, and only the other day, to my great amusement, I was talking to an influential member of the Royal Society Club about the possibility of fusing it with the Phil. Club, and he said, forgetting I was a member of the latter: "Oh! we don't want any of those wire-pullers!" Poor dear innocent dull-as-ditchwater Phil. Club!

Mention has already been made of the unveiling of the Darwin statue at South Kensington on June 9, when, as President of the Royal Society, Huxley delivered an address in the name of the Memorial Committee, on handing over the statue of Darwin to H.R.H. the Prince of Wales, as representative of the Trustees of the British Museum. The concluding words of the speech deserve quotation :—

We do not make this request [*i.e.* to accept the statue] for the mere sake of perpetuating a memory ; for so long as men occupy themselves with the pursuit of truth, the name of Darwin runs no more risk of oblivion than does that of Copernicus, or that of Harvey.

Nor, most assuredly, do we ask you to preserve the statue in its cynosural position in this entrance hall of our National Museum of Natural History as evidence that Mr. Darwin's views have received your official sanction ; for science does not recognise such sanctions, and commits suicide when it adopts a creed.

No, we beg you to cherish this memorial as a symbol by which, as generation after generation of students enter yonder door, they shall be reminded of the ideal according to which they must shape their lives, if they would turn to the best account the opportunities offered by the great institution under your charge.

Nor was this his only word about Darwin. Somewhat later, Professor Mivart sent him the proofs of an article on Darwin, asking for his criticism, and received the following reply, which describes

better than almost any other document, the nature
of the tie which united Darwin and his friends,
and incidentally touches the question of Galileo's
recantation :—

Nov. 12, 1885.

My dear Mr. Mivart—I return your proof with
many thanks for your courtesy in sending it. I fully
appreciate the good feeling shown in what you have
written, but as you ask my opinion, I had better say
frankly that my experience of Darwin is widely different
from yours as expressed in the passages marked with
pencil. I have often remarked that I never knew any
one of his intellectual rank who showed himself so
tolerant to opponents, great and small, as Darwin did.
Sensitive he was in the sense of being too ready to be
depressed by adverse comment, but I never knew any one
less easily hurt by fair criticism, or who less needed to be
soothed by those who opposed him with good reason.

I am sure I tried his patience often enough, without
ever eliciting more than a " Well there's a good deal in
what you say ; but—" and then followed something which
nine times out of ten showed he had gone deeper into the
business than I had.

I cannot agree with you, again, that the acceptance of
Darwin's views was in any way influenced by the strong
affection entertained for him by many of his friends. What
that affection really did was to lead those of his friends
who had seen good reason for his views to take much
more trouble in his defence and support, and to strike out
much harder at his adversary than they would otherwise
have done. This is pardonable if not justifiable—that
which you suggest would to my mind be neither.

I am so ignorant of what has been going on during the
last twelvemonth, that I know nothing of your controversy
with Romanes. If he is going to show the evolution of

intellect from sense, he is the man for whom I have been waiting, as Kant says.

In your paper about scientific freedom, which I read some time ago with much interest, you alluded to a book or article by Father Roberts on the Galileo business. Will you kindly send me a postcard to say where and when it was published.

I looked into the matter when I was in Italy, and I arrived at the conclusion that the Pope and the College of Cardinals had rather the best of it. It would complete the paradox if Father Roberts should help me to see the error of my ways.—Ever yours very faithfully,

T. H. HUXLEY.

August and September, as said above, were spent in England, though with little good effect. Filey was not a success for either himself or his wife. Bournemouth, where they joined their eldest daughter and her family, offered a "temperature much more to the taste of both of us," and at least undid the mischief done by the wet and cold of the north.

The mean line of health was gradually rising; it was a great relief to be free at length from administrative distractions, while the retiring pensions removed the necessity of daily toil. By nature he was like the friend whom he described as "the man to become hipped to death without incessant activity of some sort or other. I am sure that the habit of incessant work into which we all drift is as bad in its way as dram-drinking. In time you cannot be comfortable without the stimulus." But the variety of interests which filled his mind prevented him from feeling the void of inaction after a busy life. And

just as he was at the turning-point in health, he
received a fillip which started him again into vigorous
activity—the mental tonic bracing up his body and
clearing away the depression and languor which had
so long beset him.

The lively fillip came in the shape of an article in
the November *Nineteenth Century*, by Mr. Gladstone,
in which he attacked the position taken up by Dr.
Réville in his *Prolegomena to the History of Religions*,
and in particular, attempted to show that the order of
creation given in Genesis i., is supported by the
evidence of science.   This article, Huxley used
humorously to say, so stirred his bile as to set his
liver right at once; and though he denied the soft
impeachment that the ensuing fight was what had set
him up, the marvellous curative effects of a Gladstonian dose, a remedy unknown to the pharmacopœia,
became a household word among family and friends.

His own reply, "The Interpreters of Genesis and
the Interpreters of Nature," appeared in the December
number of the *Nineteenth Century* (*Collected Essays*, iv.
p. 139).   In January 1886 Mr. Gladstone responded
with his "Proem to Genesis," which was met in
February by "Mr. Gladstone and Genesis" (*Collected
Essays*, iv. p. 164).   Not only did he show that
science offers no support to the "fourfold" or the
"fivefold" or any other order obtained from Genesis
by Mr. Gladstone, but in a note appended to his
second article he gives what he takes to be the proper
sense of the "Mosaic" narrative of the Creation (iv.

p. 195), not allowing the succession of phenomena to represent an evolutionary notion, as suggested, of a progress from lower to higher in the scale of being, a notion assuredly not in the mind of the writer, but deducing this order from such ideas as, putting aside our present knowledge of nature, we may reasonably believe him to have held.

A vast subsidiary controversy sprang up in the *Times* on Biblical exegetics; where these touched him at all, as, for instance, when it was put to him whether the difference between the "Rehmes" of Genesis and "Sheh-retz" of Leviticus, both translated "creeping things," did not invalidate his argument as to the identity of such "creeping things," he had examined the point already, and surprised his interrogator, who appeared to have raised a very pretty dilemma, by promptly referring him to a well-known Hebrew commentator.

Several letters refer to this passage of arms. On December 4, he writes to Mr. Herbert Spencer :—

Do read my polishing off of the G.O.M. I am proud of it as a work of art, and as evidence that the volcano is not yet exhausted.

To Lord Farrer

4 Marlborough Place,
Dec. 6, 1885.

My dear Farrer—From a scientific point of view Gladstone's article was undoubtedly not worth powder and shot. But, on personal grounds, the perusal of it sent me blaspheming about the house with the first healthy expression of wrath known for a couple of years—to my wife's great alarm—and I should have "busted up" if I had not given vent to my indignation; and secondly, all orthodoxy was gloating over the slap in the face which the G.O.M. had administered to science in the person of Réville.

The ignorance of the so-called educated classes in this country is stupendous, and in the hands of people like Gladstone it is a political force. Since I became an official of the Royal Society, good taste seemed to me to dictate silence about matters on which there is "great division among us." But now I have recovered my freedom, and I am greatly minded to begin stirring the fire afresh.

Within the last month I have picked up wonderfully. If dear old Darwin were alive he would say it is because I have had a fight, but in truth the fight is consequence and not cause. I am infinitely relieved by getting rid of the eternal strain of the past thirty years, and hope to get some good work done yet before I die, so make ready for the part of the judicious bottle-holder which I have always found you.—Ever yours very faithfully,

T. H. HUXLEY.

4 MARLBOROUGH PLACE,
Jan. 13, 1886.

MY DEAR FARRER—My contribution to the next round was finished and sent to Knowles a week ago. I confess it to have been a work of supererogation; but the extreme shiftiness of my antagonist provoked me, and I was tempted to pin him and dissect him as an anatomico-psychological exercise. May it be accounted unto me for righteousness, though I laughed so much over the operation that I deserve no credit.

I think your notion is a very good one, and I am not

sure that I shall not try to carry it out some day.   In the meanwhile, however, I am bent upon an enterprise which I think still more important.

After I have done with the reconcilers, I will see whether theology cannot be told her place rather more plainly than she has yet been dealt with.

However, this between ourselves, I am seriously anxious to use what little stuff remains to me well, and I am not sure that I can do better service anywhere than in this line, though I don't mean to have any more controversy if I can help it.

(Don't laugh and repeat Darwin's wickedness.)—Ever yours very faithfully,   T. H. HUXLEY.

However, this "contribution to the next round" seemed to the editor rather too pungent in tone. Accordingly Huxley revised it, the letters which follow describing the process :—

4 MARLBOROUGH PLACE, N.W.,
Jan. 15, 1886.

MY DEAR KNOWLES—I will be with you at 1.30.   I spent three mortal hours this morning taming my wild cat.   He is now castrated ; his teeth are filed ; his claws are cut ; he is taught to swear like a "mieu"; and to spit like a cough ; and when he is turned out of the bag you won't know him from a tame rabbit.—Ever yours,
T. H. HUXLEY

4 MARLBOROUGH PLACE, N.W.,
Jan. 20, 1886.

MY DEAR KNOWLES—Here is the debonnaire animal finally titivated, and I quite agree, much improved, though I mourn the loss of some of the spice.   But it is an awful smash as it stands—worse than the first, I think.

I shall send you the MS. of the *Evolution of Theology* to-day or to-morrow. It will not do to divide it, as I want the reader to have an *aperçu* of the whole process from Samuel of Israel to Sammy of Oxford.

I am afraid it will make thirty or thirty-five pages, but it is really very interesting, though I say it as shouldn't.

Please have it set up in slip, though, as it is written after the manner of a judge's charge, the corrections will not be so extensive, nor the strength of language so well calculated to make a judicious editor's hair stand on end, as was the case with the enclosed (in its unregenerate state).—Ever yours very truly,   T. H. HUXLEY.

Some time later, on September 14, 1890, writing to Mr. Hyde Clarke, the philologist, who was ten years his senior, he remarks on his object in undertaking this controversy :—

I am glad to see that you are as active-minded as ever. I have no doubt there is a great deal in what you say about the origin of the myths in Genesis. But my sole point is to get the people who persist in regarding them as statements of fact to understand that they are fools.

The process is laborious, and not yet very fruitful of the desired conviction.

To Sir Joseph Prestwich

4 MARLBOROUGH PLACE, N.W.,
January 16, 1886.

MY DEAR PRESTWICH—Accept my best thanks for the volume of your Geology, which has just reached me.

I envy the vigour which has led you to tackle such a task, and I have no doubt that when I turn to your book for information I shall find reason for more envy in the thoroughness with which the task is done.

I see Mr. Gladstone has been trying to wrest your scripture to his own purposes, but it is no good.   Neither the fourfold nor the fivefold nor the sixfold order will wash.—Ever yours very faithfully,   T. H. HUXLEY.

TO PROFESSOR POULTON [1]

4 MARLBOROUGH PLACE,
Feb. 19, 1886.

DEAR MR. POULTON—I return herewith the number of the *Expositor* with many thanks.   Canon Driver's article contains as clear and candid a statement as I could wish of the position of the Pentateuchal cosmogony from his point of view.   If he more thoroughly understood the actual nature of paleontological succession—I mean the species by species replacement of old forms by new,—and if he more fully appreciated the great gulf fixed between the ideas of " creation " and of " evolution," I think he would see (1) that the Pentateuch and science are more hopelessly at variance than even he imagines, and (2) that the Pentateuchal cosmogony does not come so near the facts of the case as some other ancient cosmogonies, notably those of the old Greek philosophers.

Practically, Canon Driver, as a theologian and Hebrew scholar, gives up the physical truth of the Pentateuchal cosmogony altogether.   All the more wonderful to me, therefore, is the way in which he holds on to it as embodying theological truth.   So far as this question is concerned, on all points which can be tested, the Pentateuchal writer states that which is not true.   What, therefore, is his authority on the matter—creation by a Deity—which cannot be tested?   What sort of "inspiration" is that which leads to the promulgation of a fable as divine truth, which forces those who believe in that inspiration to hold on, like grim death, to the literal

---

[1] Hope Professor of Zoology at Oxford.

truth of the fable, which demoralises them in seeking for all sorts of sophistical shifts to bolster up the fable, and which finally is discredited and repudiated when the fable is finally proved to be a fable? If Satan had wished to devise the best means of discrediting "Revelation" he could not have done better.

Have you not forgotten to mention the leg of Archæopteryx as a characteristically bird-like structure? It is so, and it is to be recollected that at present we know nothing of the greater part of the skeletons of the older mesozoic mammals—only teeth and jaws. What the shoulder-girdle of Stereognathus might be like is uncertain.—Ever yours very faithfully,

T. H. HUXLEY.

The following letters have a curious interest as showing what, in the eyes of a supporter of educational progress, might and might not be done at Oxford to help on scientific education :—

TO THE MASTER OF BALLIOL

4 MARLBOROUGH PLACE,
Dec. 21, 1885.

MY DEAR MASTER [1]—I have been talking to some of my friends about stimulating the Royal Society to address the Universities on the subject of giving greater weight to scientific acquirements, and I find that there is a better prospect than I had hoped for of getting President and Council to move. But I am not quite sure about the course which it will be wisest for us to adopt, and I beg a little counsel on that matter.

I presume that we had better state our wishes in the

---

[1] This is from the first draft of the letter. Huxley's letters to Jowett were destroyed by Jowett's orders, together with the rest of his correspondence.

form of a letter to the Vice-Chancellor, and that we may prudently ask for the substitution of modern languages (especially German) and elementary science for some of the subjects at present required in the literary part of the examinations of the scientific and medical faculties. If we could gain this much it would be a great step, not only in itself, but in its reaction on the schools.—Ever yours very faithfully, T. H. HUXLEY.

4 MARLBOROUGH PLACE,
Dec. 26, 1885.

MY DEAR FOSTER—Please read the enclosed letter from Jowett (confidentially). I had suggested the possibility of diminishing the Greek and Latin for the science and medical people, but that, you see, he won't have. But he is prepared to load the classical people with science by way of making things fair.

It may be worth our while to go in for this, and trust to time for the other. What say you?

Merry Christmas to you. The G.O.M. is going to reply, so I am likely to have a happy New Year! I expect some fun, and I mean to make it an occasion for some good earnest.—Ever yours very faithfully,

T. H. HUXLEY.

So ends 1885, and with it closes another definite period of Huxley's life. Free from official burdens and official restraints, he was at liberty to speak out on any subject; his strength for work was less indeed, but his time was his own; there was hope that he might still recover his health for a few more years. And though the ranks of his friends were beginning to thin, though he writes (May 20, to Professor Bartholomew Price) :—

The "gaps" are terrible accompaniments of advancing

life. It is only with age that one realises the full truth of Goethe's quatrain :—

> Eine Bruche ist ein jeder Tag, etc.

and again :—

The *x* Club is going to smithereens, as if a charge of dynamite had been exploded in the midst of it. Busk is slowly fading away. Tyndall is, I fear, in a bad way, and I am very anxious about Hooker :—

still the club hung together for many years,. and outside it were other devoted friends, who would have echoed Dr. Foster's good wishes on the last day of the year :—

A Happy New Year! and many of them, and may you more and more demonstrate the folly of strangling men at sixty.

CHAPTER XVIII

1886

THE controversy with Mr. Gladstone indicates the nature of the subject that Huxley took up for the employment of his newly obtained leisure. Chequered as this leisure was all through the year by constant illness, which drove him again and again to the warmth of Bournemouth or the brisk airs of the Yorkshire moors in default of the sovereign medicine of the Alps, he managed to write two more controversial articles this year, besides a long account of the "Progress of Science," for Mr. T. Humphry Ward's book on *The Reign of Queen Victoria,* which was to celebrate the Jubilee year 1887. Examinations—for the last time, however—the meetings of the Eton Governing Body, the business of the Science Schools, the Senate of the London University, the Marine Biological Association, the Council of the Royal Society, and a round dozen of subsidiary committees, all claimed his attention. Even when driven out of town by his bad health, he would come up for a few days at a time to attend necessary meetings.

One of the few references of this period to biological research is contained in a letter to Professor Pelseneer of Ghent, a student of the Mollusca, who afterwards completed for Huxley the long unfinished monograph on "Spirula" for the *Challenger* Report.

4 MARLBOROUGH PLACE,
Jan. 8, 1886.

DEAR SIR—Accept my best thanks for the present of your publications. As you may imagine, I find that on the cretaceous crustaceans very interesting. It was a rare chance to find the branchiæ preserved.

I am glad to be able to send you a copy of my memoir on the morphology of the Mollusca. It shows signs of age outside, but I beg you to remember that it is 33 years old.

I am rejoiced to think you find it still worth consulting. It has always been my intention to return to the subject some day, and to try to justify my old conclusions—as I think they may be justified.

But it is very doubtful whether my intention will now ever be carried into effect.—I am yours very faithfully,   T. H. HUXLEY.

Mr. Gladstone's second article appeared in the January number of the *Nineteenth Century*, to this the following letter refers :—

4 MARLBOROUGH PLACE, N.W.,
Jan. 21, 1886.

MY DEAR SKELTON—Thanks for your capital bit of chaff. I took a thought and began to mend (as Burns' friend and *my* prototype (G.O.M.) is not yet recorded to

have done) about a couple of months ago, and then
Gladstone's first article caused such a flow of bile that I
have been the better for it ever since.

I need not tell you I am entirely crushed by his
reply—still the worm will turn and there is a faint
squeak (as of a rat in the mouth of a terrier) about to be
heard in the next *Nineteenth.*

But seriously, it is to me a grave thing that the
destinies of this country should at present be seriously
influenced by a man, who, whatever he may be in the
affairs of which I am no judge—is nothing but a copious
shuffler, in those which I do understand.—With best
wishes to Mrs. Skelton and yourself, ever yours very
faithfully,

T. H. HUXLEY.

With the article in the February number of the
Nineteenth Century, he concluded his tilt with Mr.
Gladstone upon the interpretation of Genesis. His
supposed "unjaded appetite" for controversy was
already satiated ; and he begged leave to retire from
"that 'atmosphere of contention' in which Mr.
Gladstone has been able to live, alert and vigorous
beyond the common race of men, as if it were purest
mountain air," for the "Elysium" of scientific debate,
which "suits my less robust constitution better."
A vain hope. Little as he liked controversy at
bottom, in spite of the skill—it must be allowed, at
times, a pleasurable skill—in using the weapons of
debate, he was not to avoid it any more than he was
to avoid the east wind when he went to Bournemouth
from early in February till the end of March, of
which he writes on February 23 :—

The "English Naples" is rather Florentine so far as a
bitter cold east wind rather below than above 0°C. goes,
but from all I hear it is a deal better than London, and
I am picking up in spite of it. I wish I were a
Holothuria, and could get on without my viscera. I
should do splendidly then.

Here he wrote a long article on the "Evolution
of Theology" (*Collected Essays*, iv. 287) which
appeared in the March and April numbers of the
Nineteenth Century. It was a positive statement of
the views he had arrived at, which underlay the very
partial—and therefore misleading—exposition of
them possible in controversy. He dealt with the
subject, not with reference to the truth or falsehood
of the notions under review, but purely as a question
of anthropology, "a department of biology to which
I have at various times given a good deal of attention." Starting with the familiar ground of the
Hebrew Scriptures, he thus explains the paleontological method he proposes to adopt :—

In the venerable record of ancient life, miscalled a
book, when it is really a library comparable to a selection
of works from English literature between the times of
Beda and those of Milton, we have the stratified deposits
(often confused and even with their natural order inverted) left by the stream of the intellectual and moral
life of Israel during many centuries. And, embedded in
these strata, there are numerous remains of forms of
thought which once lived, and which, though often unfortunately mere fragments, are of priceless value to the anthropo-
logist. Our task is to rescue these from their relatively unimportant surroundings, and by careful comparison with

existing forms of theology to make the dead world which they record live again.

A subsequent letter to Professor Lewis Campbell bears upon this essay. It was written in answer to an inquiry prompted by the comparison here drawn between the primitive spiritual theories of the books of Judges and Samuel, and the very similar development of ideas among the Tongans, as described by Mariner, who lived many years among the natives.

HODESLEA, *Oct.* 10, 1894.

MY DEAR CAMPBELL—I took a good deal of trouble years ago to satisfy myself about the point you mention, and I came to the conclusion that Mariner was eminently trustworthy, and that Martin was not only an honest, but a shrewd and rather critical, reporter. The story he tells about testing Mariner's version of King Theebaw's oration shows his frame of mind (and is very interesting otherwise in relation to oral tradition).

I have a lot of books about Polynesia, but of all I possess and have read, Mariner is to my mind the most trustworthy.

The missionaries are apt to colour everything, and they never have the chance of knowing the interior life as Mariner knew it. It was this conviction that led me to make Mariner my *cheval de bataille* in "Evolution of Theology."

I am giving a great deal of trouble—ill for the last week, and at present with a sharp lumbago! so nice! With our love to Mrs. Campbell and yourself—Ever yours,   T. H. H.

The circumstances under which the following letter was written are these. The activity of the

Home Rulers and the lethargy of Unionists had caused one side only of the great question then agitating English politics to be represented in the American press, with the result that the funds of the Nationalists were swelled by subscriptions from persons who might have acted otherwise if the arguments on the other side had been adequately laid before them.

Mr. Albert Grey, M.P., therefore had arranged for a series of clear, forcible pronouncements from strong representative Englishmen against a separate Parliament, to be cabled over to New York to a syndicate of influential newspapers, and his American advisers desired that the opening statement should be from Huxley.

Although it will be seen from the letter that he would not undertake this task, Mr. Grey showed the letter to one or two of the leading Liberal Unionists to strengthen their hands, and begged permission to publish it for the benefit of the whole party. Accordingly, it appeared in the *Times* of April 13, 1886.

CASALINI, W. BOURNEMOUTH,
March 21, 1886.

DEAR MR. GREY—I am as much opposed to the Home Rule scheme as any one can possibly be, and if I were a political man I would fight against it as long as I had any breath left in me ; but I have carefully kept out of the political field all my life, and it is too late for me now to think of entering it.

Anxious watching of the course of affairs for many years past has persuaded me that nothing short of some sharp and sweeping national misfortune will convince the majority of our countrymen that government by average opinion is merely a circuitous method of going to the devil; and that those who profess to lead but in fact slavishly follow this average opinion are simply the fastest runners and the loudest squeakers of the herd which is rushing blindly down to its destruction.

It is the electorate, and especially the Liberal electorate, which is responsible for the present state of things. It has no political education. It knows well enough that 2 and 2 won't make 5 in a ledger, and that sentimental stealing in private life is not to be tolerated; but it has not been taught the great lesson in history that there are like verities in national life, and hence it easily falls a prey to any clever and copious fallacy-monger who appeals to its great heart instead of reminding it of its weak head.

Politicians have gone on flattering and cajoling this chaos of political incompetence until the just penalty of believing their own fictions has befallen them, and the average member of Parliament is conscientiously convinced that it is his duty, not to act for his constituents to the best of his judgment, but to do exactly what they, or rather the small minority which drives them, tells him to do.

Have we a real statesman? a man of the calibre of Pitt or Burke, to say nothing of Strafford or Pym, who will stand up and tell his countrymen that this disruption of the union is nothing but a cowardly wickedness— an act bad in itself, fraught with immeasurable evil— especially to the people of Ireland; and that if it cost his political existence, or his head, for that matter, he is prepared to take any and every honest means of preventing the mischief?

I see no sign of any. And if such a man should come to the front what chance is there of his receiving loyal and continuous support from a majority of the House of Commons? I see no sign of any.

There was a time when the political madness of one
party was sure to be checked by the sanity, or at any
rate the jealousy of the other. At the last election I
should have voted for the Conservatives (for the first
time in my life) had it not been for Lord Randolph
Churchill ; but I thought that by thus jumping out of
the Gladstonian frying-pan into the Churchillian fire
I should not mend matters, so I abstained altogether.

Mr. Parnell has great qualities. For the first time
the Irish malcontents have a leader who is not eloquent,
but who is honest ; who knows what he wants and
faces the risks involved in getting it. Our poor Right
Honourable Rhetoricians are no match for this man
who understands realities. I believe also that Mr. Parnell's
success will destroy the English politicians who permit
themselves to be his instruments, as soon as bitter
experience of the consequences has brought Englishmen
and Scotchmen (and I will add Irishmen) to their senses.

I suppose one ought not to be sorry for that result,
but there are men among them over whose fall all will
lament.—I am, yours very faithfully,　T. H. HUXLEY.

Some of the newspapers took these concluding
paragraphs to imply support of Parnell, so that at
the end of June he writes :—

The *Tribune* man seems to have less intelligence than
might be expected. I spoke approvingly of the way in
which Parnell had carried out his policy, which is rather
different from approving the policy itself.

But these newspaper scribes don't take the trouble to
understand what they read.

While at Bournemouth he also finished and sent
off to the *Youth's Companion*, an American paper,
an article on the evolution of certain types of the

house, called "From the Hut to the Pantheon." Beginning with a description of the Pantheon, that characteristically Roman work with its vast dome, so strongly built that it is the only great dome remaining without a flaw :—

> For a long time (he says) I was perplexed to know what it was about the proportions of the interior of the Pantheon which gave me such a different feeling from that made by any other domed space I had ever entered.

The secret of this he finds in the broad and simple design peculiar to the building, and then shows in detail how

> the round hut, the Ædes Vestæ, and the Pantheon are so many stages in a process of architectural evolution which was effected between the first beginnings of Roman history and the Augustan age.

The relation between the beehive hut, the *terremare*, and the pile-dwellings of Italy lead to many suggestive bits of early anthropology, which, it may be hoped, bore fruit in the minds of some of his youthful readers.

We find him also reading over proofs for Mr. Herbert Spencer, who, although he might hesitate to ask for his criticism with respect to a subject on which they had a "standing difference," still

> concluded that to break through the long-standing usage, in pursuance of which I have habitually submitted my biological writing to your castigation, and so often profited by so doing, would seem like a distrust of your candour —a distrust which I cannot entertain.

So he wrote in January; and on March 19 he wrote again, with another set of proofs—

Toujours l'audace! More proofs to look over. Don't write a critical essay, only marginal notes. Perhaps you will say, like the Roman poet to the poetaster who asked him to erase any passages he did not like, and who replied, "One erasure will suffice"—perhaps you will say, "There needs only one marginal note."

To this he received answer :—

CASALINI, W. BOURNEMOUTH,
March 22, 1886.

MY DEAR SPENCER—More power to your elbow! You will find my blessing at the end of the proof.

But please look very carefully at some comments which are not merely sceptical criticisms, but deal with matters of fact.

I see the difference between us on the speculative question lies in the conception of the primitive protoplasm. I conceive it as a mechanism set going by heat—as a sort of active crystal with the capacity of giving rise to a great number of pseudomorphs; and I conceive that external conditions favour one or the other pseudomorph, but leave the fundamental mechanism untouched.

You appear to me to suppose that external conditions modify the machinery, as if by transferring a flour-mill into a forest you could make it into a saw-mill. I am too much of a sceptic to deny the possibility of anything—especially as I am now so much occupied with theology—but I don't see my way to your conclusion.

And that is all the more reason why I don't want to stop you from working it out, or rather to make the "one erasure" you suggest. For as to stopping you, "ten on me might," as the navvy said to the little special

constable who threatened to take him into custody.—Ever
yours very faithfully,   T. H. HUXLEY.

Warmth and sea-fogs here for a variety.

One more letter may be given from this time at
Bournemouth—a letter to his eldest daughter on the
loss of her infant son :—

CASALINI, W. BOURNEMOUTH,
March 2, 1886.

It's very sad to lose your child just when he was
beginning to bind himself to you, and I don't know that
it is much consolation to reflect that the longer he had
wound himself up in your heart-strings the worse the
tear would have been, which seems to have been inevitable
sooner or later.   One does not weigh and measure these
things while grief is fresh, and in my experience a deep
plunge into the waters of sorrow is the hopefullest way of
getting through them on to one's daily road of life again.
No one can help another very much in these crises of life ;
but love and sympathy count for something, and you
know, dear child, that you have these in fullest measure
from us.

On coming up to London in April he was very
busy, among other things, with a proposal that the
Marine Biological Association, of which he was
President, should urge the Government to appoint a
scientific adviser to the Fishery Board. A letter of
his on this subject had appeared in the *Times* for
March 30.   There seemed to him, with his practical
experience of official work, insuperable objections to
the status of such an officer. Above all, he would
be a representative of science in name, without any

responsibility to the body of scientific men in the country. Some of his younger colleagues on the Council, who had not enjoyed the same experience, thought that he had set aside their expressions of opinion too brusquely, and begged Sir M. Foster, as at once a close friend of his, and one to whose opinion he paid great respect, to make representations to him on their behalf, which he did in writing, being kept at home by a cold. To this letter, in which his friend begged him not to be vexed at a very plain statement of the other point of view, but to make it possible for the younger men to continue to follow his lead, he replied :—

4 MARLBOROUGH PLACE,
April 5, 1886.

MY DEAR FOSTER—Mrs. Foster is quite right in looking sharp after your colds, which is very generous of me to say, as I am down in the mouth and should have been cheered by a chat.

I am very glad to know what our younger friends are thinking about. I made up my mind to some such result of the action I have thought it necessary to take. But I have no ambition to lead, and no desire to drive them, and if we can't agree, the best way will be to go our ways separately. . . .

Heaven forbid that I should restrain anybody from expressing any opinion in the world. But it is so obvious to me that not one of our friends has the smallest notion of what administration in fishery questions means, or of the danger of creating a scientific Frankenstein in that which he is clamouring for, that I suppose I have been over-anxious to prevent mischief, and seemed domineering.

Well, I shall mend my ways.   I must be getting to be an old savage if you think it risky to write anything to me.—Ever yours,   T. H. HUXLEY.

But he did not stay long in London.   By April 20 he was off to Ilkley, where he expected to stay "for a week or two, perhaps longer."   On the 24th he writes to Sir M. Foster :—

I was beginning to get wrong before we left Bournemouth, and went steadily down after our return to London, so that I had to call in a very shrewd fellow who attends my daughter M——.   Last Monday he told me that more physicking was no good, and that I had better be off here, and see what exercise and the fresh air of the moors would do for me.   So here I came, and mean to give the place a fair trial.

I do a minimum of ten miles per diem without fatigue, and as I eat, drink, and sleep well, there ought to be nothing the matter with me.   Why, under these circumstances, I should never feel honestly cheerful, or know any other desire than that of running away and hiding myself, I don't know.   No explanation is to be found even in Foster's *Physiology!*   The only thing my demon can't stand is sharp walking, and I will give him a dose of that remedy when once I get into trim.

Indeed he was so much better even after a single day at Ilkley, that he writes home :—

It really seems to me that I am an impostor for running away, and I can hardly believe that I felt so ill and miserable four-and-twenty hours ago.

And on the 28th he writes to Sir M. Foster :—

I have been improving wonderfully in the last few days.   Yesterday I walked to Bolton Abbey, the Strid,

etc., and back, which is a matter of sixteen miles, without
being particularly tired, though the afternoon sun was as
hot as midsummer.

It is the old story—a case of candle-snuff—some
infernal compound that won't get burnt up without
more oxygenation than is to be had under ordinary conditions. . . .

I want to be back and doing something, and yet have
a notion that I should be wiser if I stopped here a few
weeks and burnt up my rubbish effectually. A good deal
will depend upon whether I can get my wife to join
me or not. She has had a world of worry lately.

As to his fortunate choice of an hotel, " I made up
my mind," he writes, "to come to this hotel merely
because Bradshaw said it was on the edge of the moor
—but for once acting on an advertisement turned out
well." The moor ran up six or seven hundred feet
just outside the garden, and the hotel itself was well
outside and above the town and the crowd of visitors.
Here, with the exception of a day or two in May, and
a fortnight at the beginning of June, he stayed till
July, living as far as possible an outdoor life, and
getting through a fair amount of correspondence.

It was not to be expected that he should long
remain unknown, and he was sometimes touched,
more often bored, by the forms which this recognition
took. Thus two days after his arrival he writes
home :—

Sitting opposite to me at the *table d'hôte* here is a
nice old Scotch lady. People have found out my name
here by this time, and yesterday she introduced herself to
me, and expressed great gratitude for the advice I gave to

a son of hers two or three years ago. I had great difficulty
in recollecting anything at all about the matter, but it
seems the youngster wanted to go to Africa, and I advised
him not to, at anyrate at present. However, the poor
fellow went, and died, and they seem to have found a
minute account of his interview with me in his diary.

But all were not of this kind. On the 26th he
writes :—

I took a three hours' walk over the moors this morning
with nothing but grouse and peewits for company, and it
was perfectly delicious. I am beginning to forget that I
have a liver, and even feel mildly disposed to the two fools
of women between whom I have to sit every meal.

27th.—. . . . I wish you would come here if only
for a few days—it would do you a world of good after
your anxiety and wear and tear for the last week. And
you say you are feeling weak. Please come and let me
take care of you a bit; I am sure the lovely air here
would set you up. I feel better than I have for
months. . . .

The country is lovely, and in a few days more all the
leaves will be out. You can almost hear them bursting.
Now come down on Saturday and rejoice the " sair een "
of your old husband who is wearying for you.

Another extract from the same correspondence
expresses his detestation for a gross breach of confidence :—

April 22.—. . . I have given Mr. —— a pretty
smart setting down for sending me Ruskin's letter to
him ! It really is iniquitous that such things should be
done. Ruskin has a right to say anything he likes in a
private letter and —— must be a perfect cad to send it
on to me.

The following letter on the ideal of a Paleontological Museum is a specialised and improved version of his earlier schemes on the same subject:—

4 MARLBOROUGH PLACE,
May 3, 1886.

MY DEAR FOSTER—I cannot find Hughes' letter, and fancy I must have destroyed it. So I cannot satisfy Newton as to the exact terms of his question.

But I am quite clear that my answer was not meant to recommend any particular course for Cambridge, when I know nothing about the particular circumstances of the case, but referred to what I should like to do if I had *carte blanche.*

It is as plain as the nose on one's face (mine is said to be very plain) that Zoological and Botanical collections should illustrate (1) Morphology, (2) Geographical Distribution, (3) Geological Succession.

It is also obvious to me that the morphological series ought to contain examples of all the extinct types in their proper places. But I think it will be no less plain to any one who has had anything to do with Geology and Paleontology that the great mass of fossils is to be most conveniently arranged stratigraphically. The Jermyn St. Museum affords an example of the stratigraphical arrangement.

I do not know that there is anywhere a collection arranged according to Provinces of Geographical Distribution. It would be a great credit to Cambridge to set the example of having one.

If I had a free hand in Cambridge or anywhere else, I should build (A) a Museum, open to the public, and containing three strictly limited and selected collections; one morphologically, one geographically, and one stratigraphically arranged; and (B) a series of annexes arranged for storage and working purposes to contain the material

which is of no use to any but specialists. I am convinced
that this is the only plan by which the wants of ordinary
people can be supplied efficiently, while ample room is
afforded for additions to any extent without large expense
in building.

On the present plan or no plan, Museums are built at
great cost, and in a few years are choked for want of
room.

If you have the opportunity, I wish you would explain that I gave no opinion as to what might or might
not be expedient under present circumstances at Cambridge. I do not want to seem meddlesome.—Ever yours
very faithfully,   T. H. HUXLEY.

Don't forget Cayley.

N.B.—As my meaning seems to have been misunderstood, I wish, if you have the chance, you would make it
clear that I do not want three brick and mortar museums
—but one public museum—containing a threefold collection of typical forms, a biological Trinity in Unity in
fact.

It might conciliate the clerics if you adopted this
illustration. But as *your own*, mind. I should not like
them to think me capable of it.

However, even Ilkley was not an infallible cure.
Thus he writes to Sir M. Foster :—

May 17.—I am ashamed of myself for not going to
town to attend the Gov. Grant Committee and Council,
but I find I had better stop here till the end of the
month, when I must return for a while anyhow.

I have improved very much here, and so long as I
take heaps of exercise every day I have nothing to complain of beyond a fit of blue devils when I wake in the
morning.

But I don't want to do any manner of work, still less

any manner of play, such as is going on in London at this time of year, and I think I am wise to keep out of it as long as I can.

I wish I knew what is the matter with me. I feel always just on the verge of becoming an absurd old hypochondriac, and as if it only wanted a touch to send me over.

May 27.—. . . . The blue devils worry me far less than they did. If there were any herd of swine here I might cast them out altogether, but I expect they would not go into blackfaced sheep.

I am disposed to stop not more than ten days in London, but to come back here and bring some work with me. In fact I do not know that I should return yet if it were not that I do not wish to miss our usual visit to Balliol, and that my Spanish daughter is coming home for a few months. . . .

I am overwhelmed at being taken at my word about scientific federation.[1] "Something will transpire" as old Gutzlaff[2] said when he flogged plaintiff, defendant and witnesses in an obscure case.

P.S.—I have had an invitation from —— to sign "without committing myself to details" an approbation of his grand scheme.[3] A stupendous array of names appear thus committed to the "principle of the Bill." I prefer to be the Hartington of the situation.

During this first stay in London he wrote twice to Mr. Herbert Spencer, from whom he had received not only some proofs, as before, on biological points, but others from his unpublished autobiography.

---

[1] *I.e.* a federation between the Royal Society and scientific societies in the colonies.

[2] This worthy appears to have been an admiral on the China station about 1840.

[3] For the reorganisation of the Fisheries Department.

After twice reading these, Huxley had merely marked a couple of paragraphs containing personal references which might possibly be objectionable "to the 'heirs, administrators and assigns,' if there are any, or to the people themselves if they are living still." He continues, June 1 :—

> You will be quite taken aback at getting a proof from me with so few criticisms, but even I am not so perverse as to think that I can improve your own story of your own life !
>
> I notice a curious thing. If Ransom [1] had not overworked himself, I should probably not be writing this letter.
>
> For if he had worked less hard I might have been first and he second at the Examination at the University of London in 1845. In which case I should have obtained the Exhibition, should not have gone into the navy, and should have forsaken science for practice. . . .

Again on June 4 :—

> MY DEAR SPENCER—Here's a screed for you ! I wish you well through it.
>
> Mind, I have no *a priori* objection to the transmission of functional modifications whatever. In fact, as I told you, I should rather like it to be true.
>
> But I argued against the assumption (with Darwin as I do with you) of the operation of a factor which, if you will forgive me for saying so, seems as far off support by trustworthy evidence now as ever it was.—Ever yours very faithfully,
>
> T. H. HUXLEY.

---

[1] Dr. Ransom of Nottingham.

On the same day he wrote to Mr., afterwards Sir John, Skelton :—

4 MARLBOROUGH PLACE, LONDON, N.W.
June 4, 1886.

MY DEAR SKELTON—A civil question deserves a civil answer—Yes. I am sorry to say I know—nobody better —" what it is to be unfit for work." I have been trying to emerge from that condition, first at Bournemouth, and then at Ilkley, for the last five months, with such small success that I find a few days in London knocks me up, and I go back to the Yorkshire moors next week.

We have no water-hens there—nothing but peewits, larks, and occasional grouse—but the air and water are of the best, and the hills quite high enough to bring one's muscles into play.

I suppose that Nebuchadnezzar was quite happy so long as he grazed and kept clear of Babylon ; if so, I can hold him for my Scripture parallel.

I wish I could accept your moral No. 2, but there is amazingly little evidence of "reverential care for un-offending creation" in the arrangements of nature, that I can discover. If our ears were sharp enough to hear all the cries of pain that are uttered in the earth by men and beasts, we should be deafened by one continuous scream !

And yet the wealth of superfluous loveliness in the world condemns pessimism. It is a hopeless riddle.—Ever yours,　　　　　　　　　　　T. H. HUXLEY.

Please remember me to Mrs. Skelton.

The election of a new Headmaster (Dr. Warre) at Eton, where he was a member of the Governing Body, was a matter of no small concern to him at this moment. Some parts of the existing system seemed impossible to alter, though a reform in the actual scheme and scope of teaching seemed to him both possible and necessary for the future well-being of the school. He writes to his eldest son on July 6, 1886 :—

> The whole system of paying the Eton masters by the profits of the boarding-houses they keep is detestable to my mind, but any attempt to alter it would be fatal.
> . . . I look to the new appointment with great anxiety. It will make or mar Eton. If the new Headmaster has the capacity to grasp the fact that the world has altered a good deal since the Eton system was invented, and if he has the sense to adapt Eton to the new state of things, without letting go that which was good in the old system, Eton may become the finest public school in the country.
> If on the contrary he is merely a vigorous representative of the old system pure and simple, the school will go to the dogs.
> I think it is not unlikely that there may be a battle in the Governing Body over the business, and that I shall be on the losing side. But I am used to that, and shall do what I think right nevertheless.

The same letter contains his reply to a suggestion that he should join a society whose object was to prevent a railway from being run right through the Lake district.

> I am not much inclined to join the "Lake District Defence Society." I value natural beauty as much as most people—indeed I value it so much, and think so highly of its influence that I would make beautiful scenery accessible to all the world, if I could. If any engineering or mining work is projected which will really destroy the beauty of the Lakes, I will certainly oppose it, but I

am not disposed, as Goschen said, to "give a blank cheque" to a Defence Society, the force of which is pretty certain to be wielded by the most irrational fanatics among its members.

Only the other day I walked the whole length of Bassenthwaite from Keswick and back, and I cannot say that the little line of rails which runs along the lake, now coming into view and now disappearing, interfered with my keen enjoyment of the beauty of the lake any more than the macadamised road did. And if it had not been for that railway I should not have been able to make Keswick my headquarters, and I should have lost my day's delight.

People's sense of beauty should be more robust. I have had apocalyptic visions looking down Oxford Street at a sunset before now.—Ever, dear lad, your loving father, T. H. HUXLEY

After this he took his wife to Harrogate, "just like Clapham Common on a great scale," where she was ordered to drink the waters. For himself, it was as good as Ilkley, seeing that he needed "nothing but fresh air and exercise, and just as much work that interests me as will keep my mind from getting 'blue mouldy.'" The work in this case was the chapter in the Life of Charles Darwin, which he had promised Mr. F. Darwin to finish before going abroad.

On July 10, he writes to Sir M. Foster on the rejection of the Home Rule Bill :—

The smashing of the G.O.M. appears to be pretty complete, though he has unfortunately enough left to give him the means of playing an ugly game of obstruction in the next Parliament.

You have taken the shine out of my exultation at

Lubbock's majority—though I confess I was disheartened to see so many educated men going in for the disruption policy.   If it were not for Randolph I should turn Tory, but that fellow will some day oust Salisbury as Dizzy ousted old Derby, and sell his party to Parnell or anybody else who makes a good bid.

We are flourishing on the whole.   Sulphide of wife joins with me in love.—Ever yours,   T. H. H.

On the 21st he writes :—

The formation of Huxley sulphide will be brought to a sudden termination to-morrow when we return to London.   The process has certainly done my wife a great deal of good and I wish it could have gone on a week or two longer, but our old arrangements are upset and we must start with the chicks for Switzerland on the 27th, that is next Tuesday.

CHAPTER XIX

1886

THE earlier start was decided upon for the sake of one of his daughters, who had been ill. He went first to Evolena, but the place did not suit him, and four days after his arrival went on to Arolla, whence he writes on August 3 :—

We reached Evolena on Thursday last. . . . We had glorious weather Thursday and Friday, and the latter day (having both been told carefully to avoid over-exertion) the wife and I strolled, quite unintentionally, as far as the Glacier de Ferpècle and back again. Luckily the wife is none the worse, and indeed, I think I was the more tired of the two. But we saw at once that Evolena was a mistake for our purpose, and were confirmed in that opinion by a deluge of rain on Saturday. The hotel is down in a hole at the tail of a dirty Swiss village, and only redeemed by very good cooking. So, Sunday being fine, I, E. and H. started up here to prospect, 18 miles up and down, and 2000 feet to climb, and did it beautifully. It is just the place for us, at the tail of a glacier in the midst of a splendid amphitheatre of 11–12000 feet snow heights, and yet not bare and waste, any quantity of stone-pines growing about. . . . I rather long for the flesh-pots of Evolena—cooking here being decidedly rudimentary—otherwise we are very well off.

The keen air of six thousand feet above sea level worked wonders with the invalids. The lassitude of the last two years was swept away, and Huxley came home eager for active life. Here too it was that, for occupation, he took up the study of gentians; the beginning of that love of his garden which was so great a delight to him in his last years. On his return home he writes :—

4 MARLBOROUGH PLACE,
Sept. 10, 1886.

MY DEAR FOSTER—We got back last evening after a very successful trip. Arolla suited us all to a T, and we are all in great force. As for me, I have not known of the existence of my liver, and except for the fact that I found fifteen or sixteen miles with a couple of thousand feet up and down quite enough, I could have deluded myself into the fond imagination that I was twenty years younger.

By way of amusement I bought a Swiss Flora in Lausanne and took to botanising—and my devotion to the gentians led the Bishop of Chichester—a dear old man, who paid us (that is the hotel) a visit—to declare that I sought the "Ur-gentian" as a kind of Holy Grail. The only interruption to our felicity was the death of a poor fellow, who was brought down on a guide's back from an expedition he ought not to have undertaken, and whom I did my best to keep alive one night. But rapid pleuritic effusion finished him the next morning, in spite of (I hope not in consequence of) such medical treatment as I could give him.

I see you had a great meeting at Birmingham, but I know not details. The delegation to Sydney is not a bad idea, but why on earth have they arranged that it shall arrive in the middle of the hot weather? Speechifying with the thermometer at 90° in the shade will try the nerves of the delegates, I can tell them.

I shall remain quietly here and see whether I can stand London. I hope I may, for the oestrus of work is upon me—for the first time this couple of years. Let me have some news of you. With our love to your wife and you—Ever yours,   T. H. HUXLEY.

4 MARLBOROUGH PLACE, N.W.,
Sept. 14, 1886.

MY DEAR DONNELLY—I hear that some of your alguazils were looking after me yesterday, so I had better give myself up at once—hoping it will be considered in the sentence.

The fact is I have been going to write to you ever since we came back last Thursday evening, but I had about fifty other letters to write and got sick of the operation.

We are all in great force, and as for me, I never expected a year ago to be as well as I am. I require to look in the glass and study the crows' feet and the increasing snow cap on the summit of my Tête noire (as it once was), to convince myself I am not twenty years younger.

How long it will last I don't feel sure, but I am going to give London as little chance as possible.

I trust you have all been thriving to a like extent. Scott[1] wrote to me the other day wanting to take his advanced flock (2—one, I believe, a ewe-lamb) to Kew. I told him I had no objection, but he had better consult you.

I have not been to S.K. yet—as I have a devil (botanical—) and must satisfy him before doing anything

---

[1] Assistant Professor of Botany at the Royal College of Science.

else. It's the greatest sign of amendment that I have
gone in for science afresh. When I am ill (and consequently venomous), nothing satisfies me but gnawing at
theology ; it's a sort of crib-biting.

Our love to Mrs. Donnelly. I suppose G.H.[1] is by
this time a kind of Daniel Lambert physically and
Solomon mentally—my blessing to him.—Ever yours
very faithfully,   T. H. HUXLEY.

As a sequel to the sad event mentioned in the
former letter, the relations of the young man who
had died so suddenly at Arolla wished to offer Huxley
some gift in grateful recognition of the kindness he
had shown to the poor fellow ; but being unable to
fix upon any suitable object, begged him to accept a
considerable sum of money and expend it on any
object he pleased as a memento.   To this he replied,
November 21, 1886 :—

I am very much obliged for the kindly recognition of
my unfortunately unavailing efforts to be of service to
your brother-in-law which is contained in your letter.

But I and those who right willingly helped me did
nothing more than our plain duty in such a case ; and
though I fully appreciate the motives which actuate
Mrs. —— and yourself and friends, and would gladly
accept any trifle as a memento of my poor friend (I call
him so, for we really struck up a great friendship in our
twelve hours' acquaintance), I could not with any comfort
use the very handsome cheque you offer.

Let me propose a compromise. As you will see by
the enclosed paper, a colleague of mine has just died
leaving widow and children in very poor circumstances.

---

[1] Gordon Huxley Donnelly, Sir John's son.

Contribute something to the fund which is being raised for their benefit, and I shall consider it as the most agreeable present you could possibly make to me.

And if you wish me to have a personal memento of our friend, send me a pipe that belonged to him. I am greatly devoted to tobacco, and will put it in a place of honour in my battery of pipes.

The bracing effects of Arolla enabled him to stay two months in town before again retiring to Ilkley to be "screwed up." He had on the stocks his Gentian Paper and the chapter for the Darwin Life, besides the chapter on the Progress of Science for the *Reign of Queen Victoria*, all of which he finished off this autumn; he was busy with Technical Education, and the Egyptian borings which were being carried out under the superintendence of the Royal Society. Finally he was induced by a "diabolical plot" on the part of Mr. Spencer to read, and in consequence to answer, an article in the *Fortnightly* for November by Mr. Lilly on "Materialism and Morality." These are the chief points with which the following correspondence is concerned.

4 MARLBOROUGH PLACE,
Sept. 16, 1886.

MY DEAR FOSTER—I enclose the Report [1] and have nothing to suggest except a quibble at p. 4. If you take a stick in your hand you may feel lots of things and determine their form, etc., with the other end of it, but surely the stick is properly said to be insensible. D°.

---

[1] The Annual Report of the Examiners in Physiology under the Science and Art Department, which, being still an Examiner he had to sign.

with the teeth. I feel very well with mine (which are paid for) but they are surely not sensible? Old Tomes once published the opinion that the contents of the dentine tubules were sensory nerves, on the ground of our feeling so distinctly through our teeth. He forgot the blind man's stick. Indeed the reference of sensation to the end of a stick is one of the most interesting of psychological facts.

It is extraordinary how those dogs of examinees return to their vomit. Almost all the obstinate fictions you mention are of a quarter of a century date. Only then they were dominant and epidemic—now they are sporadic.

I wish Pasteur or somebody would find some microbe with which the rising generation could be protected against them.

We shall have to re-arrange the Examination business —this partner having made his fortune and retiring from firm. Think over what is to be done.—Ever yours,

T. H. H.

You don't happen to grow gentians in your Alpine region, do you?

Of his formal responsibility for the examinations he had written earlier in the year:—

WELLS HOUSE, ILKLEY,
June 15, 1886.

MY DEAR DONNELLY—I think it is just as well that you could not lay your hands on ink, for if you had you would only have blacked them. (*N.B.* This is a goak.)

You know we resolved that it was as well that I should go on as Examiner (unpaid) this year. But I rather repent me of it—for although I could be of use over the questions, I have had nothing to do with checking the results of the Examination except in honours, and I suspect that Foster's young Cambridge allies tend always to screw the standard up.

I am inclined to think that I had much better be out of it next year. The attempt to look over examination papers now would reduce the little brains I have left to mere pulp—and, on the other hand, if there is any row about results, it is not desirable that I should have to say that I have not seen the answers.

When I go you will probably get seven devils worse than the first—but that is not the fault of the first devil.

I am picking up here wonderfully in spite of the bad weather. It rained hard yesterday and blew ditto—to-day it is blowing dittoes—but there is sunshine between the rain and squalls.

I hope you are better off. What an outlandish name "Tetronila." I don't believe you have spelt it right. With best regards to Mrs. Donnelly and my godson—
Ever yours, T. H. HUXLEY.

4 MARLBOROUGH PLACE,
Sept. 16, 1886.

MY DEAR HOOKER—I have sucked Grisebach's brains, looked up *Flora B. Americana*, and *F. Antarctica and New Zealand*, and picked about in other quarters. I found I knew as much as Grisebach had to tell me (and more) about *lutea, purpureo-punctata, acaulis, campestris,* and the *verna* lot, which are all I got hold of at Arolla. But he is very good in all but classification, which is logically "without form and void, and darkness on the face of it."

I shall have to verify lots of statements about gentians I have not seen, but at present the general results are very curious and interesting. The species fall into four groups, one *primary* least differentiated—three, specialised.

1. Lobes of corolla fringed. 2. Coronate. 3. Interlobate (*i.e.* not the "plica" between the proper petals).

Now the interesting point is that the Antarctic species are all primary and so are the great majority of the Andean forms. *Lutea* is the only old-world primary, unless the Himalayan *Moorcroftiana* belongs here. The Arctic forms are also primary, but the petals more extensively united.

The specialised types are all Arctogeal with the exception of half a dozen or so Andean species including *prostrata.*

There is a strange general parallelism with the crayfishes ! which also have their primary forms in Australia and New Zealand, avoid E. S. America and Africa, and become most differentiated in Arctogæa. But there are also differences in detail.

It strikes me that this is uncommonly interesting; but, of course, all the information about the structure of the flowers, etc., I get at second hand, wants verifying.

Have you done the gentians of your *Flora Indica* yet ? Do look at them from this point of view.

I cannot make out what Grisebach means by his division of Chondrophylla. What is a "cartilaginous" margin to a leaf ?—" Folia margine *cartilaginea !* " He has a lot of Indian sp. under this head.

I send you a rough scheme I have drawn up. Please let me have it back. Any annotations thankfully received. Shan't apologise for bothering you.

I hope the pension is settled at last.—Ever yours,

T. H. HUXLEY.

4 MARLBOROUGH PLACE,
Sept. 22, 1886.

MY DEAR HOOKER—I have written to Lubbock a long screed stating my views [1] with unmistakable distinctness as politeful as may be, and asking him, if he thought well, to send them on to whomsoever it may concern. As old Gutzlaff [2] used to say when he wanted to get

---

[1] Referring to the relations between the S. K. department and the City and Guilds Committee on Technical Education.
[2] See p. 451.

evidence from a Chinee—"Gif him four dozen, someting vill transpire." At any rate the Chinee transpired, and I hope some official will.

Here beginneth more gentian craze.

I have not examined *Moorcroft.* yet, but if the figure in Roxb. is trustworthy it's a primary and no mistake. I can't understand your admitting *Amarellae* without coronae. The presence of a corona is part of the definition of the *amarella* group, and an *amarella* without a corona is a primary *ipso facto.*

Taking the facts as I have got them in the rough, and subject to minor verifications, the contrast between the Andean, Himalayan, and Caucasian Gentian Florae is very striking.

	Simplices.	Ciliatae.	Coronatae.	Interlobatae.
Andes	27	0 (?)	15	2
Himalayas	1 (*Moorcroft.*)	0	4	32
Caucasus Pyrenees (all one)	2 (*lutea umbellata*)	2	5	21

I don't think *Ciliatae* worth anything as a division. I took it as it stood.

It is clear that migration helps nothing, as between the old-world and S. American Florae. It is the case of the Tapirs (Andean and Sino-Malayan) over again. Relics of a tertiary Flora which once extended from S. America to Eurasia through N. America (by the west, probably).

I see a book by Engler on the development of Floras since tertiary epoch. Probably the beggar has the idea. —Ever yours, T. H. HUXLEY.

GODALMING, *Sept.* 25, 1886.

MY DEAR FOSTER—We are here till to-morrow on a visit to Leonard, seeing how the young folks keep house.

I brought the Egyptian report down with me. It is

very important, and in itself justifies the expenditure.
Any day next (that is to say this) week that you like I
can see Col. Turner.   If you and Evans can arrange a
day I don't think we need mind the rest of the Committee.   We must get at least two other borings ten or
fifteen miles off, if possible on the same parallel, by hook
or by crook.   It will tell us more about the Nile valley
than has ever been known.   That Italian fellow who
published sections must have lied considerably.

Touching gentians, I have not examined your
specimen yet, but it certainly did not look like *Andrewsii*.
You talk of having *acaulis* in your garden.   That is one
of the species I worked out most carefully at Arolla, but
its flowering time was almost over, and I only got two
full-blown specimens to work at.   If you have any in
flower and don't mind sacrificing one with a bit of the
rhizoma, and would put it in spirit for me, I could
settle one or two points still wanting.   Whisky will do,
and you will be all the better for not drinking the
whisky !

The distributional facts, when you work them in
connection with morphology, are lovely.   We put up
with Donnelly on our way here.   He has taken a cottage
at Felday, eleven miles from hence, in lovely country—
on lease.   I shall have to set up a country residence
some day, but as all my friends declare their own locality
best, I find a decision hard.   And it is a bore to be tied
to one place.—Ever yours,   T. H. HUXLEY.

4 MARLBOROUGH PLACE,
Oct. 20, 1886.

MY DEAR HOOKER—I wish you would not mind the
trouble of looking through the enclosed chapter which I
have written at F. Darwin's request, and tell me what
you think of it.   F. D. thinks I am hard upon the
"Quarterly Article," but I read it afresh and it is
absolutely scandalous.   The anonymous vilifiers of the

present day will be none the worse for being reminded that they may yet hang in chains. . . .

It occurs to me that it might' be well to add a paragraph or two about the two chief objections made formerly and now to Darwin, the one, that it is introducing "chance" as a factor in nature, and the other that it is atheistic.

Both assertions are utter bosh. None but parsons believe in "chance"; and the philosophical difficulties of Theism now are neither greater nor less than they have been ever since Theism was invented.—Ever yours,

T. H. H.

The following letter to Mr. Edmund Gosse, who, just before, had been roughly handled in the *Quarterly Review*, doubtless owed some of its vigour to these newly revived memories of the *Quarterly* attack on Darwin. But while the interest of the letter lies in a general question of literary ethics, the proper methods and limits of anonymous criticism, it must be noted that in this particular case its edge was turned by the fact that immediately afterwards, the critic proceeded to support his criticisms elsewhere under his own name :—

Oct. 22, 1886.

DEAR SIR—I beg leave to offer you my best thanks for your letter to the *Athenæum*, which I have just read, and to congratulate you on the force and completeness of your answer to your assailant.

It is rarely worth while to notice criticism, but when a good chance of exposing one of these anonymous libellers who disgrace literature occurs, it is a public duty to avail oneself of it.

Oddly enough, I have recently been performing a

similar "haute œuvre." The most violent, base, and ignorant of all the attacks on Darwin at the time of the publication of the "Origin of Species" appeared in the *Quarterly Review* of that time; and I have built the reviewer a gibbet as high as Haman's.

All good men and true should combine to stop this system of literary moonlighting.—I am yours very faithfully, T. H. HUXLEY.

On the same date appeared his letter to the *Pall Mall Gazette*, which was occasioned by the perversion of the new Chair of English Literature at Oxford to "Middle English" philology :—

I fully agree with you that the relation of our Universities to the study of English literature is a matter of great public importance; and I have more than once taken occasion to express my conviction—Firstly, that the works of our great English writers are pre-eminently worthy of being systematically studied in our schools and universities as literature; and secondly, that the establishment of professional chairs of philology, under the name of literature, may be a profit to science, but is really a fraud practised upon letters.

That a young Englishman may be turned out of one of our universities, "epopt and perfect," so far as their system takes him, and yet ignorant of the noble literature which has grown up in those islands during the last three centuries, no less than of the development of the philosophical and political ideas which have most profoundly influenced modern civilisation, is a fact in the history of the nineteenth century which the twentieth will find hard to believe; though, perhaps, it is not more incredible than our current superstition that whoso wishes to write and speak English well should mould his style after the models furnished by classical antiquity. For my part, I venture to doubt the

wisdom of attempting to mould one's style by any other
process than that of striving after the clear and forcible
expression of definite conceptions ; in which process the
Glassian precept, "first catch your definite conceptions,"
is probably the most difficult to obey. But still I mark
among distinguished contemporary speakers and writers
of English, saturated with antiquity, not a few to whom,
it seems to me, the study of Hobbes might have taught
dignity ; 'of Swift, concision and clearness ; of Goldsmith
and Defoe, simplicity.

Well, among a hundred young men whose university
career is finished, is there one whose attention has ever
been directed by his literary instructors to a page of
Hobbes, or Swift, or Goldsmith, or Defoe ? In my boyhood we were familiar with *Robinson Crusoe, The Vicar
of Wakefield,* and *Gulliver's Travels ;* and though the
mysteries of "Middle English" were hidden from us,
my impression is we ran less chance of learning to write
and speak the "middling English" of popular orators
and headmasters than if we had been perfect in such
mysteries and ignorant of those three masterpieces. It
has been the fashion to decry the eighteenth century, as
young fops laugh at their fathers. But we were there in
germ ; and a "Professor of Eighteenth Century History
and Literature" who knew his business might tell young
Englishmen more of that which it is profoundly important they should know, but which at present remains
hidden from them, than any other instructor ; and,
incidentally, they would learn to know good English
when they see or hear it—perhaps even to discriminate
between slipshod copiousness and true eloquence, and that
alone would be a great gain.

As for the incitement to answer Mr. Lilly, Mr.
Spencer writes from Brighton on November 3 :—

I have no doubt your combative instincts have been
stirred within you as you read Mr. Lilly's article,

"Materialism and Morality," in which you and I are dealt with after the ordinary fashion popular with the theologians, who practically say, "You *shall* be materialists whether you like it or not." I should not be sorry if you yielded to those promptings of your combative instinct. Now that you are a man of leisure there is no reason why you should not undertake any amount of fighting, providing always that you can find foemen worthy of your steel.

I remember that last year you found intellectual warfare good for your health, so I have no qualms of conscience in making the suggestion.

To this he replies on the 7th :—

Your stimulation of my combative instincts is downright wicked. I will not look at the *Fortnightly* article lest I succumb to temptation. At least not yet. The truth is that these cursed irons of mine, that have always given me so much trouble, will put themselves in the fire, when I am not thinking about them. There are three or four already.

On November 21 Mr. Spencer sends him more proofs of his autobiography, dealing with his early life :—

See what it is to be known as an omnivorous reader —you get no mercy shown you. A man who is ready for anything, from a fairy tale to a volume of metaphysics, is naturally one who will make nothing of a fragment of a friend's autobiography !

To this he replies on the 25th :—

4 MARLBOROUGH PLACE,
Nov. 25, 1886.

MY DEAR SPENCER—In spite of all prohibition I must write to you about two things. First, as to the proof returned herewith—I really have no criticisms to make (miracles, after all, may not be incredible). I have read your account of your boyhood with great interest, and I find nothing there which does not contribute to the understanding of the man. No doubt about the truth of evolution in your own case.

Another point which has interested me immensely is the curious similarity to many recollections of my own boyish nature which I find, especially in the matter of demanding a reason for things and having no respect for authority.

But I was more docile, and could remember anything I had a mind to learn, whether it was rational or ir-rational, only in the latter case I hadn't the mind.

But you were infinitely better off than I in the matter of education. I had two years of a Pandemonium of a school (between 8 and 10) and after that neither help nor sympathy in any intellectual direction till I reached manhood. Good heavens ! if I had had a father and uncle who troubled themselves about my education as yours did about your training, I might say as Bethell said of his possibilities had he come under Jowett, "There is no knowing to what eminence I might not have attained." Your account of them gives me the impression that they were remarkable persons. Men of that force of character, if they had been less wise and self-restrained, would have played the deuce with the abnormal chicken hatched among them.

The second matter is that your diabolical plot against Lilly has succeeded—*vide* the next number of the *Fortnightly*.[1] I was fool enough to read his article, and the rest followed. But I do not think I should have troubled myself if the opportunity had not been good for clearing off a lot of old scores.

The bad weather for the last ten days has shown me

---

[1] Science and Morals, *Coll. Ess.* ix. 117.

that I want screwing up, and I am off to Ilkley on Saturday for a week or two. Ilkley Wells House will be my address. I should like to know that you are picking up again.—Ever yours very faithfully,

T. H. HUXLEY.

And again on December 13 :—

I am very glad to have news of you which on the whole is not unsatisfactory. Your conclusion as to the doctors is one I don't mind telling you in confidence I arrived at some time ago. . . .

I am glad you liked my treatment of Mr. Lilly. . . . I quite agree with you that the thing was worth doing for the sake of the public.

I have in hand another bottle of the same vintage about Modern Realism and the abuse of the word Law, suggested by a report I read the other day of one of Liddon's sermons.[1]

The nonsense these great divines talk when they venture to meddle with science is really appalling.

Don't be alarmed about the history of Victorian science.[2] I am happily limited to the length of a review article or thereabouts, and it is (I am happy to say it is nearly done) more of an essay on the history of science, bringing out the broad features of the contrast between past and present, than the history itself. It seemed to me that this was the only way of dealing with such a subject in a book intended for the general public.

The article "Science and Morals" was not only a satisfaction to himself, but a success with the readers of the *Fortnightly*. To his wife he writes :—

December 2.—Have you had the *Fortnightly*? How does my painting of the Lilly look?

---

[1] "Pseudo-Scientific Realism," *Coll. Ess.* iv. 59.   [2] See p. 461.

December 8.—Harris . . . says that my article "simply made the December number," which pretty piece of gratitude means a lively sense of favours to come.

December 13.—I had a letter from Spencer yesterday chuckling over the success of his setting me on Lilly.

Ilkley had a wonderful effect upon him. "It is quite absurd," he writes after 24 hours there, "but I am wonderfully better already." His regimen was of the simplest, save perhaps on one point. "Clark told me," he says with the utmost gravity, "always to drink tea and eat hot cake at 4.30. I have persevered, however against my will, and last night had no dreams, but slept like a top." Two hours' writing in the morning were followed by two hours' sharp walking; in the afternoon he first took two hours' walking or strolling if the weather were decent; "then Clark's prescription diligently taken" (*i.e.* tea and a pipe) and a couple of hours more writing; after dinner reading and to bed before eleven.

I am working away (he writes) in a leisurely comfortable manner at my chapter for Ward's Jubilee book, and have got the first few pages done, which is always my greatest trouble.

December 8.—. . . Canon Milman wrote to me to come to the opening of the New Buildings for Sion College, which the Prince is going to preside over on the 15th. I had half a mind to accept, if only for the drollery of finding myself among a solemn convocation of the city clergy. However, I thought it would be opening the floodgates, and I prudently declined.

One more letter may perhaps be quoted as illustrating the clearness of vision in administrative matters which made it impossible for him to sit quietly by and see a tactical blunder being committed, even though his formal position might not seem to warrant his interference. This is his *apologia* for such a step.

Dec. 16, 1886.

My DEAR FOSTER—On thinking over this morning's Committee work,[1] it strikes my conscience that being neither President or Chairman nor officer I took command of the boat in a way that was hardly justifiable.

But it occurred to me that our sagacious —— for once was going astray and playing into ——'s hands, without clearly seeing what he was doing, and I bethought me of "salus Societatis suprema lex," and made up my mind to stop the muddle we were getting into at all costs. I hope he was not disgusted nor you either. X. ought to have cut in, but he did not seem inclined to do so.

I am clearly convinced it was the right thing to do—anyhow.—Ever yours,   T. H. H.

The chronicle of the year may fitly close with a letter from Ilkley to Dr. Dohrn, *apropos* of his recommendation of a candidate for a biological professorship. The "honest sixpence got by hard labour," refers to a tour in the Highlands which he had once taken with Dr. Dohrn, when, on a rough day, they were being rowed across Loch Leven to Mary Stuart's castle. The boatman, unable to make head

---

[1] Some Committee of the Royal Society.

single-handed against the wind, asked them each to take an oar; but when they landed and Huxley tendered the fare, the honest fellow gave him back two sixpences, saying, "I canna tak' it: you have wrocht as hard as I." Each took a coin; and Huxley remarked that this was the first sixpence he had earned by manual labour. Dr. Dohrn, I believe, still carries his sixpence in memory of the occasion.

WELLS HOUSE, ILKLEY, YORKSHIRE,
Dec. 1, 1886.

MY DEAR DOHRN—You see by my address that I am *en retraite*, for a time. As good catholics withdraw from the world now and then for the sake of their souls—so I, for the sake of my body (and chiefly of my liver) have retired for a fortnight or so to the Yorkshire moors—the nearest place to London where I can find dry air 1500 feet above the sea, and the sort of uphill exercise which routs out all the unoxygenated crannies of my organism. Hard frost has set in, and I had a walk over the moorland which would have made all the blood of the Ost-see pirates—which I doubt not you have inherited—alive, and cleared off the fumes of that detestable Capua to which you are condemned. I should like to have seen the nose of one [of] your Neapolitan nobilissimes after half-an-hour's exposure to the north wind, clear and sharp as a razor, which very likely looked down on Loch Leven a few hours ago.

Ah well! "fuimus"—I am amused at the difficulty you find in taking up the position of a "grave and reverend senior"; because I can by no means accustom myself to the like dignity. In spite of my grey hairs "age hath not cooled the Douglas blood" altogether, and I have a gratifying sense that (liver permitting) I am still capable of much folly. All this, however, has not much to do with poor Dr. —— to whom, I am sorry to say, your letter could do no good, as it arrived after my colleagues and I had settled the business.

But there were a number of strong candidates who had not much chance. If it is open to me to serve him hereafter, however, your letter will be of use to him, for I know you do not recommend men lightly.

After some eighteen months of misery—the first thing that did me any good was coming here. But I was completely set up by six or seven weeks at Arolla in the Valais. The hotel was 6400 feet up, and the wife and daughters and I spent most of our time in scrambling about the 2000 feet between that and the snow. Six months ago I had made up my mind to be an invalid, but at Arolla I walked as well as I did when you and I made pilgrimages—and earned the only honest sixpence (I, at any rate) ever got for hard labour. Three months in London brought me down again, so I came here to be "mended."

You know English literature so well that perhaps you have read Wordsworth's "White Doe of Rylstone." I am in that country, within walk of Bolton Abbey.

Please remember me very kindly to the Signora—and thank her for copying the letter in such a charmingly legible hand. I wish mine were like it.

If I am alive we shall go to Arolla next summer. Could we not meet there? It is a fair half-way.—Ever yours, T. H. HUXLEY.

END OF VOL. II

Printed by R. & R. CLARK, LIMITED, Edinburgh.

Milton Keynes UK
Ingram Content Group UK Ltd.
UKHW032320161024
449665UK00001B/32

9 781108 040464